Design and the Digital Divide

Insights from 40 Years in Computer Support for Older
and Disabled People

Synthesis Lectures on Assistive, Rehabilitative, and Health-Preserving Technologies

Editor
Ronald M. Baecker, *University of Toronto*

Advances in medicine allow us to live longer, despite the assaults on our bodies from war, environmental damage, and natural disasters. The result is that many of us survive for years or decades with increasing difficulties in tasks such as seeing, hearing, moving, planning, remembering, and communicating.

This series provides current state-of-the-art overviews of key topics in the burgeoning field of assistive technologies. We take a broad view of this field, giving attention not only to prosthetics that compensate for impaired capabilities, but to methods for rehabilitating or restoring function, as well as protective interventions that enable individuals to be healthy for longer periods of time throughout the lifespan. Our emphasis is in the role of information and communications technologies in prosthetics, rehabilitation, and disease prevention.

Design and the Digital Divide: Insights from 40 Years in Computer Support for Older and Disabled People
Alan F. Newell
2011

Design and the Digital Divide: Insights from 40 Years in Computer Support for Older and Disabled People

Alan F. Newell

ISBN: 978-3-031-00464-3   paperback
ISBN: 978-3-031-01592-2   ebook

DOI 10.1007/978-3-031-01592-2

A Publication in the Springer series
SYNTHESIS LECTURES ON ASSISTIVE, REHABILITATIVE, AND HEALTH-PRESERVING TECHNOLOGIES

Lecture #1
Series Editor: Ronald M. Baecker, *University of Toronto*
Series ISSN
Synthesis Lectures on Assistive, Rehabilitative, and Health-Preserving Technologies
ISSN pending.

Design and the Digital Divide

Insights from 40 Years in Computer Support for Older and Disabled People

Alan F. Newell
Dundee University

SYNTHESIS LECTURES ON ASSISTIVE, REHABILITATIVE, AND HEALTH-PRESERVING TECHNOLOGIES #1

ABSTRACT

Demographic trends and increasing support costs means that good design for older and disabled people is an economic necessity, as well as a moral imperative.

Alan Newell has been described as "a visionary who stretches the imagination of all of us" and "truly ahead of his time". This monograph describes research ranging from developing communication systems for non-speaking and hearing-impaired people to technology to support older people, and addresses the particular challenges older people have with much modern technology.

Alan recounts the insights gained from this research journey, and recommends a philosophy, and design practices, to reduce the "Digital Divide" between users of information technology and those who are excluded by the poor design of many current systems.

How to create and lead interdisciplinary teams, and the practical and ethical challenges of working in clinically related fields are discussed. The concepts of "Ordinary and Extra-ordinary HCI", "User Sensitive Inclusive Design", and "Design for Dynamic Diversity", and the use of "Creative Design" techniques are suggested as extensions of "User Centered" and "Universal Design". Also described are the use of professional theatre and other methods for raising designers' awareness of the challenges faced by older and disabled people, ways of engaging with these groups, and of ascertaining what they "want" rather than just what they "need".

This monograph will give all Human Computer Interaction (HCI) practitioners and designers of both mainstream and specialized IT equipment much food for thought.

KEYWORDS

digital inclusion, Human Computer Interaction, HCI design, older and disabled people, augmentative and alternative communication, user sensitive inclusive design, ordinary and extra-ordinary HCI , research methodology, requirements gathering, awareness raising, historical

To Maggie

Contents

Acknowledgments

The author would like to express his gratitude to the following.

To the funding agencies and other bodies who funded the research: Engineering and Physical Sciences Research Council, National Research and Development Corporation, Scottish Higher Education Funding Council, Scottish Home and Health Department, Tayside Regional Council, Department for Trade and Industry, Department for Education and Skills, NCR (Dundee), Digital Equipment Corporation Europe (and Dave Lawrence), House of Commons, European Commission, Independent Broadcasting Authority, British Broadcasting Corporation, Intel, Palantype Organisation, , Winston Churchill Travel Trust, Leverhulme Trust, Alzheimer's Association of America, Spastics Society, Whitetop Foundation, TSB Foundation, and Private Donors.

To Barbara Johnson, Louisa Cross, and their administration team, and Derek Brankin and his technical support team.

To my academic colleagues who contributed to this research, including: Norman Alm, John Arnott, Alistair Cairns, Peter Gregor, Jesse Hoey, Janet Hughes, Stephen McKenna, Iain Murray, Adrian Pickering, Graham Pullin, Ian Ricketts, Annalu Waller, Vicki Hanson, Marianne Hickey, and Robin King.

To my research colleagues, including: Rob Baker, Andrew Lambourne, Colin Brookes, Andrew Downton, Alex Carmichael, Graeme Coleman, Anna Dickinson, Lorna Gibson, Paula Forbes, David Sloan, Bernadette Brophy-Arnott, Linda Booth, Liz Broumley, Hilary Hood, Richard Dye, Bill Beattie, Mike Smith, Andrea Szymkowiak, Greg Filz, Andy McKinley, Maggie Morgan (of Foxtrot Theatre and MM Training), Jon Brumfitt, Stefan Langer, Helen Peddie, Mark Rice, Phillipa Riley, Andrew Swiffin, Dave O'Mara, Andrew Wright, Alan McGregor, and Isla Beard (Palantypist).

To my national and international colleagues, including: Arleen Kraat, Penny Parnes, Rick Foulds. Gregg Vanderheiden, Thijs Soede, Pat Demsco, Eric Hollnagel, Alex Mihailidis, Sherri Hunnicutt, Janice Light, Peter Hutt, Leela Damodaran, Stephen Brewster, Steven von Tetzchner, David Goodall (of Soundsmove), Mike Smith, and Oli Mival.

To the Universities of Southampton and Dundee, and Principles Hamlin, Graham Bryce and Sir Alan Langlands.

To Her Majesty Queen Elizabeth the Queen Mother and Dr. Mary Dowager Countess of Strathmore, and Baron (Jack) Ashley of Stoke for their support for our research.

To licensees: Possum Controls, Lander Software, and Don Johnson, Mayer Johnson Co.

To Norman Alm, Maggie Newell, Gregg Vanderheiden, and Mickey Milner for reading and commenting on drafts of this book.

Alan F. Newell
June 2011

Foreword

Creativity researchers have found that persistence pays off and that fresh perspectives yield insights. Alan Newell's 40-year professional career confirms these findings, and in addition makes other compelling claims. Readers of his thoughtful and compelling historical review will also learn about some further claims about creativity, such as: empathy triggers innovations and theater changes minds. I was charmed by these claims, which were embedded in Newell's anecdote-filled history of his contributions.

Professional memoirs are valuable, not only to those whose work is mentioned and those whose research is aligned with the author's, but also as a guide to young researchers in related fields. The lessons of a lifetime are a precious gift to readers who are trying to decide about careers, choose research domains, or respond to their passions about a specific project. Reading about how Newell came to be the champion for disabled, older, and extraordinary users of technology should give strong encouragement for those who are worried about blazing their own trail.

The encouragement to "take a fresh perspective" is easy to give, but harder to put to work in a meaningful way. Newell's examples repeatedly show readers how to open their eyes and look at problems in fresh ways. The historic photos of early innovations are important historical records, which help readers see how new ideas developed. But getting new ideas is only the starting point in developing a successful innovation.

Newell reminds readers that "unless there is significant opposition to an idea it cannot be radical enough to have the potential to make substantial improvement." This wise advice should propel young researchers to fly into the wind, because that is the best way to gain altitude. But there is more to research success than being bold. Newell supports the common belief that teamwork is a catalyst for success, but he cautions that teams must have certain components or properties:

- Engineers with empathy, insight and imagination,

- Clinicians and other relevant disciplines,

- An understanding of the lives of the people for whom one is designing

- An appropriate research methodology,

- A knowledge of the literature, and an awareness of current commercially available devices,

- A healthy skepticism of current practice,

- Ample time thinking what they want to achieve, rather than how they are going to achieve it.

Newell's ethics are infused with human aspirations and personal connections. He believes that researchers should "have frequent contact with potential users not only as 'experimental subjects' but also as people and colleagues, to improve empathy and the intuitive skills of the researcher." Here again Newell evokes empathy as a key ingredient for success, but how many students are taught its many flavors and satisfactions in their academic training?

I've come away from reading this book with fresh perspectives about research, clarity about teamwork, and reminders about persistence. Newell's writings and his life work defined and legitimized research on extraordinary users, producing plentiful benefits for all technology users. He deserves ample recognition for catalyzing innovative research, initiating academic courses, and launching valuable products. I hope readers will experience the same warm feelings I have of respect and appreciation for my delightfully creative colleague.

Ben Shneiderman
June, 2011

CHAPTER 1

40 years–Highlights and a Brief Review

THE DIGITAL DIVIDE

Digital technology has maked an enormous difference in the lives of a great many people, but significant numbers of people have been excluded, or have excluded themselves, from these benefits. These include many older and disabled people and other minority groups including people from certain cultures and those with very poor educational achievements. This has been called the Digital Divide. As government and commerce increasingly rely on the internet, these groups are becoming more and more disenfranchised.

In the later part of the first decade of the 2000s, a range of initiatives were introduced to encourage such people to use Information Technology (IT) including advertising, provision of broadband in homes, and educational courses. A major problem, however, is that most information technology has not been designed with these groups of people in mind. They are either scared of it and/or do not think they will ever be able to use it. In order to play their part in reducing the Digital Divide, designers need take on the challenge of including the requirements of these "extra-ordinary" people in their practices.

IT systems have been developed to support older and disabled people for many years, and this book contains an historic review of research in some areas in the field. One important lesson from this research is that rules and guidelines for including disadvantaged people in designs are sometimes less than adequate.

> *Orthodoxy is the Grave of Intelligence.*
> **(Bertrand Russell)**

This book highlights insights and lessons learnt from over 40 years leading a group researching into and developing computer systems for older and disabled people. With the goal of influencing the mind sets of designers, it focuses on ideas on which to ponder, rather than describing an exhaustive research methodology.

1.1   AN INTERDISCIPLINARY EDUCATION

I was very fortunate whilst at University to have been, more or less, forced into what turned out to be an interdisciplinary education. I was too young to be trained as a computer engineer but I did learn to touch type before going up to University—in the vain hope that I would type out my lecture notes each evening. This early decision has been immensely helpful to me as a software engineer, and, in more recent times, for any writing tasks. (All students learning to touch type would, I believe, make a significant improvement to their efficiency—but in the UK this is not happening).

I had been very much focussed on science and engineering at school and read Electrical Engineering at Birmingham University. This was a traditional course, and was at the time (1959-1962) when transistors were gradually being introduced into such courses. So, I was taught about power transmission, valves (vacuum tubes), and transistors, but not computers or integrated circuits. On being awarded a BSc in 1962, I was offered a PhD studentship in that Department. My supervisor (Prof. Jack Allanson) was interested in how sound was reproduced on the cortex. He persuaded me to perform some experiments which involved playing sounds to human subjects and noting their responses. Thus, I had to learn about experimental psychology, both experimental methods and statistical analysis of results.

It had been agreed that, at the half-way point of my PhD, I would return to more traditional electrical engineering. Another staff member, however, had different ideas. He wanted some experiments done on how people recognised sonar images, and it was made clear to me that it would fit in better with a three-year PhD—as I was already up to speed on perceptual experiments. I was disappointed, but agreed to go down that path which resulted in my PhD being entirely in "Subjective Pattern Recognition", with little or no engineering.

> *Everything might be for the best in this the best of all possible worlds.*
> **(Pangloss revised: Voltaire)**

At the end of my student days I was a trained electrical engineer and, by happenstance, a self-taught experimental psychologist. Such an interdisciplinary background was unusual in those days. This combination, however, proved to be very beneficial in my career, and, towards the end of it, the concept of "inter-disciplinary research" had become very popular. So my reluctant decision proved to be a very good one—my first example of an apparent set-back turning out to be for the best in the long run.

1.2   INDUSTRY—SPEECH RECOGNITION RESEARCH

In 1965 my background in people, sound and engineering secured me a job in an industrial research laboratory that was funded to develop a speech recognition system. This was at a time when most researchers saw speech recognition as a relatively simple technical problem which "would be solved in two years, and make much money in five years". This view of speech recognition continued throughout the rest of the last century but commercial success was very elusive. Interestingly, however, some of the earliest examples of commercial success of speech technology (synthesis and recognition) were as aids for people with disabilities. I was, and remain, very conscious of the complexity of speech and of the advantages of speech recognition—except in limited and carefully controlled situations [Newell, A., 1992a]. I was in a research environment whose aims did not fit with my views of what could be achieved, and the contribution that I, and the group I was part of, were able to make to that very overly-ambitious project was not great.

There were, however, many advantages: the project purchased one of the first laboratory computers—a Digital Equipment PDP8. This had 4,096 12-bit words of storage, a keyboard and a 10-character per second printer, paper tape reader and punch, and led to me obtaining on-the-job training in software engineering. As a laboratory tool, programming in assembly language was essential, and many hours were spent trying to produce very efficient programmes which fitted into the 4K of storage that was available. Thus, I was again fortunate in being an early user of laboratory computers—initially for developing laboratory tools and simulating electronic circuits and latterly as prototype electronic systems in their own right.

The other major advantage of my time in these laboratories was that I was able to read widely in the general area of speech and to be an early investigator of human computer interfaces. Up until that time the majority of work on human factors had been supported by the military with a focus on knobs and dials. As will be seen, the background that this gave me proved to be essential to many of the research issues that I subsequently investigated.

> *Opportunism—a valid research strategy.*

What, in retrospect, was the turning point in my career path occurred by chance. It was becoming increasingly clear that the automatic speech recognition project—like the vast majority of such projects at that time—was not going to achieve its aim. One day, my immediate boss, almost as an aside, suggested that our technology might be beneficial to people with disabilities. As it was not possible at that time to recognise speech, I developed "VOTEM", a Voice Operated Typewriter employing Morse Code. The idea being that a disabled person could speak Morse Code (dots and dashes), to spell out what they wanted to type.

> *Technology push is sometimes a useful driver.*

VOTEM was licensed to a manufacturer of systems for disabled people but it was never made commercially available. This particular piece of research did not have a commercial outcome, but it did interest me in developing systems for disabled people. There are many and varied reasons why researchers move into this field—some because of a relationship either with disabled people directly, or via discussions with clinicians, but in my case it was because of a potential use for technology. Technologically led research, particularly in this field, can be problematic and, although disabled people were not involved in the development of VOTEM, they should have been. This experience gave me an interest in, and a particular perspective on, this area of research, and much of my subsequent work has promoted the idea of "user-centred design".

1.3   SOUTHAMPTON UNIVERSITY—A DEVELOPING FOCUS ON "AIDS FOR DISABLED PEOPLE"

I was appointed to a lectureship in Electronics at Southampton University and decided that developing systems for disabled people would be one of my major research interests. In comparison to the organisation of many universities in the 21st Century, the choice of research area was entirely left up to me—there was no institutional pressure for or against such a choice. The research in this, and subsequent sections of this chapter, will be described very briefly, with the details and rationale being expanded upon in later chapters.

Designing VOTEM had opened my eyes to the communication problems of people without speech, and my readings in psychology had made me aware of the many and varied characteristics of speech, including speech being more that just the words spoken, and the importance of body language. I was struck by the fact that all the systems that had been developed for non-speaking people required the non-speaking person and their communication partner to look at a single screen or printer—which meant that eye contact, which is very important in face-to-face communication between speaking people, was not possible. The "Talking Brooch" [Newell, A., 1974a] consisted of a small "rolling" alphanumeric display worn on the lapel and operated via a hand-held keyboard, and was designed to provide eye contact for non-speaking people. I wrote a simulation of this on a PDP12 (the successor to the PDP8 as a laboratory computer), and my team subsequently developed a dedicated electronic version. (In the early 1980's, even "small computers" were very large.)

Encourage serendipity.

The Talking Brooch was a major factor in my being awarded a Winston Churchill Travel Fellowship in 1976 to visit researchers in the U.S. This was an immensely useful experience, enabling me to visit the major players in the field. Many of the people I met have remained friends and colleagues throughout my career. The Winston Churchill Travel Fellowship was excellent in that its modus operandi was to choose Fellows on the basis of their ideas and then give them freedom to plan their fellowship without having to check back to the Trust. For example, I was advised not to have a full diary so that I could follow up leads that came up during the tour. If it wasn't for this ability I would never have met the New York-based speech pathologist, Arleen Kraat. She was not on my original itinerary, but became a very important mentor and supporter throughout my research activities in this field.

The Travel Fellowship confirmed my view that research into, and development of, systems to assist people with disabilities was an area which was satisfying and which I would enjoy. The area of assisting human communication was particularly interesting and held some exciting technological challenges, but serendipity led to the specific projects I and my team pursued.

1.4   A CHANCE MEETING WITH A MEMBER OF THE BRITISH PARLIAMENT

Lewis Carter-Jones, a Member of Parliament was visiting the Department at Southampton and I demonstrated the Talking Brooch to him. He was a friend of Jack (now Lord) Ashley, a labour MP who had suddenly become deaf. He suggested that the Brooch would be useful to assist Jack in the House of Commons, where he had great difficulty in following debates, and he arranged for me to meet Jack in the House. This led to my team developing a transcription system for machine shorthand, in particular the Palantype Machine that had been developed in the UK. This system provided a verbatim transcript of speech on a display screen. It became the first computer system to be used in the Chamber of the House of Commons, and led the field in commercially available real-time systems for stenograph transcription.

1.5   TELEVISION SUBTITLING

My investigations into the needs of deaf people led me to consider television subtitling, and we investigated ways in which "closed captions" could be transmitted. This research was superseded when the UK text services of Oracle and Ceefax were developed. We thus refocussed our research on the characteristics required for effective captioning, and developing equipment that would enable captioners to work efficiently. Although the Independent Television authorities supported the former research, they did not see any need for new equipment. So again there was no support from the potential users of such a system, but Andrew Lambourne, a research student at the time, continued this development both as a PhD topic and as a commercial venture [Lambourne et al., 1982a]. In 2011 he continues to run a successful company marketing this type of equipment.

Research without stakeholder support can still be valuable.

The Palantype and the Subtitling projects were brought together in our research into live subtitling. ITV supported our work, our system being used for the Charles & Diana royal wedding, whereas the BBC supported Earnest Edmunds at Leicester University. In those early days, although subtitling for deaf people added only 1/3 of one percent to the cost of programmes, it was deemed to be too high a price to pay (to assist 10% of the audience!). After much lobbying, this view changed and a large percentage of programmes in the UK are now subtitled, including 100% of the British Broadcasting Corporation's output news being subtitled mainly by stenographers.

1.6   DUNDEE UNIVERSITY

In 1980, I moved to the NCR chair of Electronics and Microcomputer Systems in Dundee University's Electrical Engineering and Electronics Department. There I founded a group investigating the uses of microcomputers with a special interest in disabled people. This again was not a "strategic" decision by the University—their aim was to expand their research and teaching in microcomputers. My research group was not a good fit with the Electrical Engineering and Electronics Department and, in 1986, the group joined Mathematics to produce a Department of Mathematics and Computer Science. Later, it became a stand-alone Department of Applied Computing and subsequently the School of Computing. My move from being essentially an electrical/electronic engineering academic to the head of an Applied Computing Department was not entirely for academic reasons, but proved to be exactly the right move to make both from teaching and research standpoints.

1.7   THE SCHOOL OF COMPUTING

A major thrust of the School of Computing at Dundee University is computer systems for areas of high social impact. It had four research groups: Assistive and Healthcare technologies, Interactive Systems design, Computational systems and the Space Technology group. There was cross-fertilization between all four groups, but, in particular, there was close links between Assistive and Health care technologies and Interactive system design. The Assistive and Healthcare technologies group contains over 30 researchers developing computer and communication technology for older and disabled people, and has become the largest and one of the most influential academic groups in the world in this field [Newell, A., 2004].

In the 1980s and 1990s, much of the group's research to support disabled people focussed on non-speaking people and the development of Augmentative and Alternative Communication (AAC) systems [Gregor et al., 1999]. These are computer systems that control a speech synthesiser—the most well-known user of such a system being Professor Steven Hawking of Cambridge University. My academic colleagues Adrian Pickering, John Arnott, Norman Alm, Annalu Waller and Ian Ricketts, plus a large number of Research Students, Assistants and Fellows worked in this field. This work which will be described in more detail in later chapters, but was aimed at increasing the rate at which non-speaking people could talk, and was based on prediction, and the use of conversational models. It is vital that such work be interdisciplinary, and our research group has included a wide variety of disciplines, including psychologists, speech and occupational therapists, linguists, philosophers, nurses, school teachers, and creative designers, as well as computer engineers and human computer interface (HCI) specialists. I, and other colleagues, also have interdisciplinary academic backgrounds, which have proved particular helpful in our research. We were concerned with providing systems that allowed the non-speaking people to transmit their personalities, rather than simply deliver messages, and the work included Iain Murray's pioneering work on inserting emotion into the output of speech synthesisers.

Our work in the AAC field led us to be considered mavericks by many speech therapists in the international field, some of whom were strongly opposed to our ideas, but Arlene Kraat was a constant source of support for our work.

> *A maverick: "a person pursuing rebellious, even potentially disruptive, policies or ideas".*

Thanks to Lynda Booth, a special education teacher, we were the first research group to show that predictive systems can assist people with spelling and language dysfunction, and John Arnott's research into disambiguation was one of the triggers for the development of the T9 system which is available in most mobile phones today. Other work the group did during this period included Peter Gregor's research into computer-supported interviewing, where we worked with child psychiatry units and collaborated with researchers in a secure mental hospital. Peter Gregor also conducted ground-breaking research into support for people with dyslexia.

> *Unusual perspectives can be valuable.*

In addition to research focussed on the needs of disabled people, we examined the wider ramifications of such work, and developed the idea of "ordinary" and "extra-ordinary" human computer interaction. The concept was that "ordinary" people operating in an "extra-ordinary" environment (high workload and stress, such as flying planes and warfare), provided similar HCI challenges to those of "extra-ordinary" (disabled) people operating in an "ordinary" environment (e.g., word processing). This led to a number of projects, and was also used to encourage researchers in the international HCI community to consider the needs of disabled people in their research. Messages from other researchers to the HCI and Design communities were focussed on the concepts of "Inclusive Design", "Design for All", and "Universal Design". In the UK at least, however, these concepts only began to be seen in mainstream Information Technology research and development in the early years of this century. The growing importance of the digital economy and the demographics of the user base gave these ideas a commercial impetus. Policy makers are realising that older and disabled people, and other minority groups, are much less represented in cyber space than young, able-bodied, educated people. This is important for the UK Government as many of the services they offer and hope to computerise are targeted at these groups.

1.8   IT SUPPORT FOR OLDER PEOPLE—THE QUEEN MOTHER RESEARCH CENTRE

In the latter part of the 20th Century, most of the research work in this field, including that at Dundee, had been focussed on disabled people. By the turn of the century, however, the wider public were beginning to become aware of the changing demographics: the world was becoming older, people were living longer, and, in the developed world, birth-rates were reducing. This led to serious concerns about how the world could support such a changing demographic, and there was a growing interest in how technology could help support older people. We realised that the characteristics of older people, some of whom had disabilities, were very different to those of the traditional disabled person for whom most computer-based technology had been developed. We thus decided to investigate how technology could support older people, and developed the concept of a research centre focussed on information technology to support older people.

> *"I don't skate to where the puck is. I skate to where the puck is going to be".*
> **Wayne Gretzky (Canadian hockey player)**

Through the good offices of Dr. Mary, Dowager Countess of Strathmore, we were able to persuade Queen Elizabeth, the Queen Mother, to give her name to this venture (see Figure 1.1), and the University to provide a purpose-designed building for the School of Computing that contained both a studio theatre and a User Centre—a facility specifically for our older users. In 2006, the Princess Royal formally opened the Queen Mother Building which housed the School (see Figure 1.2), including the research specifically aimed at supporting older and disabled users [Newell, A., 2006]. The research in the School can be seen at:
`www.computing.dundee.ac.uk`.

During this period, Peter Gregor developed links with the School of Design, and introduced an "Interactive Media Design Degree" (now called Digital Interaction Design). This led to the recruitment of Graham Pullin from the design house IDEO, who had a background in rehabilitation engineering and creative design. He has been developing the use of creative design techniques within disability and AAC research, and this has produced novel and very interesting research directions. Peter Gregor also initiated the School's Digital Media Access Group. This combines research activities with an audit and advisory service for accessibility and usability of websites and software interfaces, usability research with older and disabled people, and advice on accessibility strategies. This combination of research and service is not usual within an academic environment, but in our case has proved extremely successful, with much synergy existing between the two remits.

Norman Alm re-aligned his research from communication aids for speech-impaired people into a focus on how information technology could be used to support people with dementia. This led to the development of a multi-media-based system to encourage reminiscence and other systems to improve the quality of life of people with dementia. John Arnott and Nick Hine expanded their

CLARENCE HOUSE
SW1A 1BA

 I am sorry that I am unable to be with you this
evening but I would like to thank Baroness Greengross
for agreeing to hold this Reception at the House of
Lords.

 I am very pleased that this project to establish
the Queen Mother Centenary Research Centre for
Information Technology to Support Older People has been
launched so successfully, and I know that this is in no
small way due to the hard work of Professor Alan Newell
and his team at the University of Dundee.

 I hope that the Centre will provide new
opportunities for collaborative research and be of
benefit to many elderly and disabled people throughout
the United Kingdom. I send my best wishes for the
continued success of this endeavour.

                    ELIZABETH R
                    Queen Mother

February 2002

Figure 1.1: Letter from Her Majesty Queen Elizabeth the Queen Mother.

Figure 1.2: The Queen Mother Building.

research into Smart Housing, and Steve McKenna's research on computer vision included similar application areas.

During this time, I was particularly focussed on how the needs of older people could be included within current design methodologies such as "user-centred design". Our UTOPIA (Usable Technology for Older People Inclusive and Appropriate) project—in collaboration with Glasgow, Napier and Abertay Universities—investigated the most effective ways of including older people in the design process. We were also commissioned to act as consultants to Fujitsu, who were developing a portal for older people for the Department for Education. The research in this project formed the basis for the government's "MyGuide" website for older people. The lesson from this project was that we had to look for very powerful ways of raising software and human interface designers' awareness of characteristics of older people: not only their sensory and motor characteristics, but also the effects of their up-bringing and their relationship with new technologies.

1.9   THEATRE FOR AWARENESS RAISING AND REQUIREMENTS GATHERING

It was essential that powerful communication techniques be used for awareness raising, and we thus began a fruitful collaboration with Maggie Morgan, a script writer and theatre director (then artistic director of the Foxtrot Theatre Company and now of MM Training) who had a background in Forum

Theatre and the use of Interactive Theatre in training and community consultation, and a film maker David Goodall (of Soundsmove). In collaboration with these, and other theatre professionals, we used theatre both for requirements gathering for projects to support older people and for raising designers' awareness of the characteristics, needs and wants of older people. A particular use of theatre within a research framework was part of the Inclusive Design Educational Network of academic and industrial researchers that was aimed at producing a research agenda for this field. In parallel with developing a research agenda, this group briefed a film maker, who illustrated the research agenda developed by this project by a narrative film. "Relatively PC" and other examples of the use of theatre in this field can be seen at:

`www.computing.dundee.ac.uk/acprojects/iden`.

In 2006, I became an Emeritus Professor, and, in 2009, Vicki Hanson, formerly the Accessibility Manager of IBM, based in New York, joined the School as a full professor. Together with other research projects, she is a Principal Investigator for the Newcastle and Dundee Universities £13M Inclusive Digital Economy Hub funded by the Engineering and Physical Sciences Research Council. The School thus continues to go from strength to strength in this field of research.

1.10   PUTTING ONE'S FAITH IN STORIES

A vital part of research is communicating results—both to fellow researchers and to the public, and I have investigated ways of trying to increase the impact of one's results. I have come to the conclusion that, although data is vital, the power of the message in the data can be greatly increased by presenting a story, if possible with a personal narrative. Such stories are often denigrated as "simply anecdotes", but a good story—particularly one with some humor and/or conflict—which effectively illustrates the message behind the data, can be a very powerful tool.

1.11   SUMMARY

> *"Hold fast to dreams.*
> *For if dreams die,*
> *Life is a broken winged bird*
> *Which cannot fly".*
> **(anon.)**

This monograph essentially follows an historical perspective of research in the fields in which researchers at Dundee University's School of Computing were operating, together with lessons learned from this journey.

Chapters 2–7 focus on research into the use of information technology to improve human-to-human communication.

Chapter 2 highlights the development of VOTEM—the speech-operated typewriter for physically disabled people, the Palantype Transcription System for deaf people, and the Talking Brooch, the first system we developed for non-speaking people. Chapter 3 describes the work on television subtitling for hearing-impaired people. This includes research into both the style of captioning, and the development of efficient captioning systems for both pre-prepared and real-time captions.

The Talking Brooch led to sustained research in the AAC field. This included word prediction (Chapter 4), and its importance not only for reducing key-strokes, but also for improving literacy. Other software for dyslexia is also described in this chapter, as is research that led to disambiguation techniques. Systems based on this research are now included in most mobile phone texting systems: possibly the most ubiquitous example of research for disabled people leading to a mainstream product.

Even with word prediction, AAC devices still produced speech at very slow rates, thus other speed increasing techniques were developed. Chapter 5 examines the influence of Conversational Analysis research on the AAC field. This led to the concept of reusable conversation, and an examination of the relative importance of pragmatics and semantics in conversation. A number of systems based on this concept are described, together with the conflicts that arose from these developments.

The ultimate instantiation of "reusable conversation" is stories: these are very important for personal development and building relationships. Because of their length, however, AAC users rarely relate stories. Chapter 6 describes systems that were developed to encourage non-speaking people to relate stories.

Chapter 7 relates the lessons that the team at Dundee learned from this research and recommends ways in which the efficacy of research in this particular field can be increased.

Towards the end of the 20th Century, demographic trends clearly indicated an increasing percentage of older people in the population. People were living to an increasingly old age, and were exhibiting physical, sensory and cognitive decline. In general, information technology had not designed for such people and a "digital divide" was developing between younger people and the majority of those over 65. Chapter 8 describes a range of computer-based systems that have been designed particularly for older people, including those with dementia. Although older people may have disabilities, their general characteristics were different from those of younger disabled people. Chapter 9 describes these differences and how they relate to design challenges in Information technology and Human Computer Interface research. Chapter 9 lays out a range of methods for including older people in the design process.

The differences between older and disabled people and young able-bodied people, however, are less than might at first appear to be the case—particularly in relation to the design of IT products. Chapter 10 introduces the concept of Ordinary and Extra-Ordinary Human Computer Interaction as a way of bridging this gap, and of encouraging mainstream designers to consider the needs and wants of currently excluded populations of users. Chapter 11 suggests "User Sensitive Inclusive Design" as an expansion of "User-Centered Design" to include disadvantaged users. The philosophy behind this is described and a number of suggestions made as to how this can be implemented, and how the concept of Designing for Dynamic Diversity provides a framework for interface design that responds to the changes in users as they grow older.

It is vital for designers and users to relate to each other, and this is particularly important where there are age and cultural differences between these two groups. There is evidence that design rules and guidelines are not sufficient, and Chapter 12 describes ways in which professional theater can be used both for requirements gathering and for raising designers' awareness of the challenges older people have with technology.

Chapter 13 draws these ideas together with recommendations for design practice in the field of developing computer systems to support older and disabled people. It suggests how such approaches can benefit all users, young and old, fit and unfit, healthy or unhealthy, and with varying degrees of cognitive functioning. It recommends this approach to designers of mainstream as well as assistive technology.

CHAPTER 2

Communication Systems for Non-Speaking and Hearing-Impaired People

The development of a voice-operated typewriter for non-speaking, physically disabled people described in Chapter 1 led to the development of the Talking Brooch. This was one of the first truly portable communication aid for non-speaking people. Demonstrating this to a chance visitor to the Department introduced the challenge of providing a communication aid for a profoundly deaf Member of Parliament and led to the development of a system based on the automatic transcription of machine shorthand.

2.1 A VOICE-OPERATED TYPEWRITER FOR PHYSICALLY DISABLED PEOPLE

In my readings related to speech recognition research, I had come across a paper that was trying to automatically recognise hand-sent Morse Code. This had not been particularly successful as the timing of hand-sent Morse Code is not accurate. Fast operators can send Morse that can be understood by a human being, but differences in the lengths of the dots and dashes and the spaces within and between characters defeated automatic recognition methods.

It seemed to me, however, that, if speed was not the overarching objective, an operator could be trained to send Morse Code which could be automatically decoded. Also, the system would provide excellent feedback from errors as, if an "i" (dot-dot), was recognised as an "m" (dash-dash), the operator would know that s/he had to reduce the length of the dots. Thus, spoken Morse code was a possible way in which people who were paralysed from the neck down could type. I simulated VOTEM (Voice Operated Typewriter Employing Morse-code) on the PDP 8 to prove that this was possible and subsequently designed and built an electronic version [Newell and Nabavi, 1969, Newell, A., 1970].

Clearly, disabled people would have preferred to talk to a typewriter—as some 30-odd years later they would be able to do—but it was not possible at the time. VOTEM was an example of reducing the requirements to match what was possible. Clearly, spoken Morse code was not a viable input method for someone who could use a typewriter keyboard, but was a candidate for someone who could not. This situation is still the case: speech recognition is only really viable in situations where it is not possible or inconvenient to use a keyboard.

VOTEM—the Voice Operated Typewriter Employing Morse Code sparked my interest in developing technology to assist people with disabilities. It also introduced me to technologically assisted human-human communication. The knowledge and background I had gained in my research into Automatic Speech Recognition showed me that speech communication is very much more than the words which are spoken. Human communication is the very basis of our humanity and is a very complex and subtle process. Our communication with other human beings is not just a set of messages that we relay to other people: it is, in a very real way, our personality. If we are to develop artificial means to replace speech, we must be as concerned about the form of the communication as the efficiency of it as a message carrier.

I thus embarked on background reading in the area of what was to become known as Augmentative and Alternative Communication (AAC)—technology to support people with impaired speech and language. I also became familiar with a range of (relatively unrelated) research topics that would prove to be very useful in my future research.

At this time, Possum Controls was one of the leading developers of systems for severely paralyzed people, these were based essentially on scanning a matrix by sucking and blowing down a tube. Figure 2.1 shows an early version of such a system. This technology had been developed by Reg Mailing, who was a "visitor" at Stoke Manderville Hospital. He observed patients using a whistle to communicate with people. At that time even simple electronics was too expensive for this application, but he realized that the Strowger equipment (a two-dimensional mechanical selector mechanism used in telephone exchanges at time) was inexpensive, and could be modified to provide a scanning matrix which could control domestic equipment or an electric typewriter via a pneumatic tube. He formed the POSSUM Company [Mailing and Clarkson, 1963, Mailing, R., 1968], which in the early 21st Century is still marketing communication aids for disabled people.

By the early 1970s, a number of similar systems had begun to appear [Copeland, K., 1974, Foulds et al., 1975, Ridgeway and Mears, 1985, Vanderheiden, G., 2002]. Some examples of these are shown in Figure 2.2. There were no portable systems, and they all required the disabled person and the conversational partner to look at a remote display or printout. This meant that eye contact, and the ability to notice facial expression, which I believe is very important in face-to-face communication, was not possible. In addition, I thought that an AAC system should be instantly available—so that users did not feel that they had to wait for something important to say, before switching their system on, and that, like speech, it should provide a transitory communication—not a "permanent" written one. An example of the dangers of a printed output was related to me by Arlene Kraat. A non-speaking patient had printed out the message "you did not brush my hair properly" to a nurse, but, instead of this being interpreted as a relatively unimportant comment, it was taken as a formal complaint. This example highlights the difference between the impact of spoken and written messages, which also became a concern in our research into television sub-titling.

Figure 2.1: An early Possum system.

2.2   THE TALKING BROOCH—A COMMUNICATION AID FOR NON-SPEAKING PEOPLE

A challenge for AAC systems was to create a portable device that was mounted near to the face. Conventional displays were not appropriate as visual displays at that time were heavy, large, and expensive. My "eureka" moment occurred whilst I was travelling through King's Cross station where there was a rolling newscaster display, and I remembered Taenzer's [1970] work aimed at improving the "Opticon". This was a reading aid for the blind, in which the operator scanned a printed page via an array of small vibrators on their finger.

Taenzer had shown that a rolling display of only one character width was readable. In a preliminary experiment we showed that the reading speed increased as the number of characters displayed increased [Newell et al., 1975]. We thus conducted formal reading experiments using a simulated display between 1 and 12 characters long. We also compared rolling and walking displays

(a)

(b)

(c)

Figure 2.2: Early AAC Devices. (a) Portaprinter - commercially available; (b) TIC - developed by Rick Foulds, Tufts University, Boston; (c) AutoCom - developed by Greg Vanderheiden, University of Wisconsin-Maddison.

(where the location of the character matrixes were fixed and the letters jumped from one matrix to the next). Users performed significantly better with the rolling display and, with a 5-character display, 99% of sentences could be read at a (fast typing) rate of 60 wpm [Newell and Brumfitt, 1979b].

Although a rolling display would be more expensive to produce, we decided that this was essential, and, as a compromise between cost and readability, we built a 5-character prototype using individual light-emitting diodes. This can been seen in Figure 2.3(a). A later version shown in Figure 2.3(b) used light emitting diode array. The whole system consisted of the display mounted in a breast pocket, a battery pack and a keyboard. This fulfilled the requirements I had laid down, and an indication of the success of the Talking Brooch idea was given by a child's parent who said that the very first time he had told a joke was via his Talking Brooch.

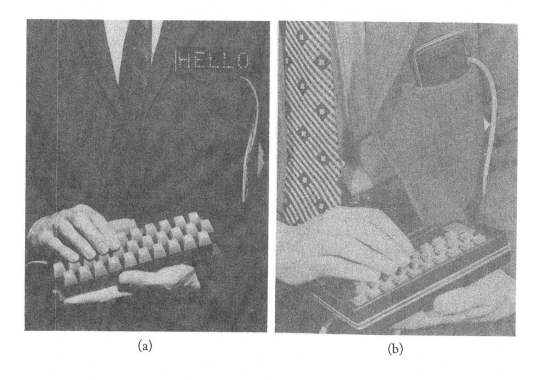

(a)   (b)

Figure 2.3: The Talking Brooch. (a) A prototype Talking Brooch; (b) the commercially available Talking Brooch.

A portable device the ELKOMI 2 marketed by Diode (Amsterdam) had a 9-letter walking display, but, even though the display was longer, our results would indicate that it would be less easy to read at normal typing speeds.

At much the same time Toby Churchill of Toby Churchill Ltd had developed the Lightwriter, which consisted of a much longer single line display integrated with a keyboard [Lowe et al., 1974], and is shown in Figure 2.4(a). In later versions of the Lightwriter, such as that shown in Figure 2.4(b), there is a two-sided display—one side facing the communication partner, and another identical display facing the operator—again with an integrated keyboard. Although the Lightwriter was not

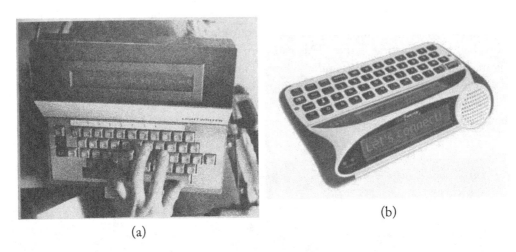

(a)

(b)

Figure 2.4: The Lightwriter. (a) Earliest version; (b) 2010 version.

as good at promoting eye contact, it did facilitate an appropriate body language for face-to-face communication. In addition, the integrated nature of the system meant that there was only one "box", and no external wiring. The only other portable device available at that time was the Cannon Communicator, which essentially was similar in style to a pocket calculator, but with an alphanumeric keyboard and a strip printer.

In 1976, Vanderheiden [1976] reviewed the literature in this field, addressing the issues of accessing communication aids and the relative merits of direct selection (as employed in the Talking Brooch) and scanning and encoding techniques. He also surveyed the range of AAC devices that were available at that time. There were very few portable devices but, in addition to the ones mentioned above, he described the MCM device marketed by Micon Industries (California) that had been primarily designed as a communication for the deaf. He also cited the Versicom and Autocom, developed by the Trace Centre at the University of Wisconsin-Madison, as examples of wheelchair portable systems.

The Lightwriter and the Talking Brooch had made slightly different design compromises. The Talking Brooch majored on eye contact and immediacy, whereas the Lightwriter allowed the disabled user to see what they were typing, and had no external wiring or sockets, with their associated fragility. The Cannon Communicator had the advantages of a single box, but did not facilitate appropriate body language. The Talking Brooch [Newell, A., 1974a] was marketed by the University of Southampton, and had modest sales. The Lightwriter is still selling well in the early 21st Century. This shows how important it is to really examine the use of any system in real contexts and, where necessary, to compromise on the "purity" of the goal for pragmatic reasons.

A full appreciation of the potential uses of systems in real contexts is essential.

The experience of developing the Talking Brooch led to a range of projects all designed to improve the efficacy of communication aids for people with speech and language dysfunction. It also led to my award of a Winston Churchill Travel Fellowship to investigate communication aids for non-speaking people in the U.S. This formed the basis of much of my future work in this field. I met Arlene Kraat, who subsequently became my mentor from the field of Speech Therapy, and President of the International Society of Augmentative and Alterative Communication. Other very important friends and colleagues from that Fellowship included Greg Vanderheiden, from the University of Wisconsin Madison—who has made a major contribution to technological development for disabled people at a research and development and political levels—and Rick Foulds, who led very exciting research in this area for many years at the Universities of Tufts and Delaware.

2.3   SPEECH TRANSCRIPTION FOR DEAF PEOPLE

I had noted that the Talking Brooch could also be used for deaf people, but the catalyst for my next research projects was a visit to the Department of Lewis Carter Jones, MP. He was a colleague of Jack (now Lord) Ashley [Ashley, J., 1973] who had become deaf and was struggling to continue his parliamentary career. It is impossible to lip read in the Chamber, and he was surviving by relying on a fellow MP, sitting next to him in the House, writing notes for him on what was said. I arranged to meet Jack and his wife Pauline in the House to demonstrate the Talking Brooch. His (accurate) assessment was that it would be no better than written notes—what he required was a verbatim transcript of what was being said. A good typist can type at 60-80 words per minute, but speech can reach over 200 words per minute. In the British Parliament, particularly at Prime Minister's Questions, it often happened that an innocent aside (which would not be deemed worth writing down for Jack) would be picked up a couple of speeches later—often as a joke. If he did not have a verbatim transcript, Jack was likely to miss the point of these references [Ashley, J., 1992]. This was similar to the reported complaints of deaf students who were offered a real-time version of lectures on a visual system using an operator who listened to the lecture and dictated a synopsis to a typist [Hales, G., 1976].

Fortune favors the prepared mind.
(Pasteur 1854)
Therefore be a research "butterfly" and read widely.

It was clear to me that automatic speech recognition would not work within this environment: a couple of years previously I had written that "we must put firmly out of our minds any thoughts of, or hopes for, a 'mechanical typist'. If we do this we will be in a better position to specify the sort of machine that can be built and may be useful in helping the deaf" [Newell, A., 1974b]. (The limitations of speech recognition are discussed in more detail in Newell [Newell, A., 1992c]). This was my opportunity to put these comments into effect. During my ASR research I had come across attempts, some ten years previously, to transcribe the British Palantype machine shorthand [Price, W., 1971], and the American machine shorthand system, Stenograph [Newitt and Odarchenko, 1970]. I thus knew that it was possible to input Palantype data into a computer, but also that current systems required large computers and were not accurate enough to make them a commercial possibility for Court Reporting. Jack Ashley, however, did not need a correct transcription just one which was readable, but he did need a portable system which had to work in real-time. Thus research which had been a commercial failure at that time led my team to develop a system that was appropriate for people with disabilities.

> **The excellent is an enemy of the good.**

Palantype, Stenograph, and the French Grand Jean system work in similar ways (Figure 2.5(a)) shows a Palantype Machine). All these systems have chord keyboards, where a number of keys are pressed at the same time, and they work in a syllabic mode, which means that each syllable is encoded in one stroke in a pseudo phonetic form. The left-hand keys being used to encode the initial phoneme, the right-hand keys encoding the final phoneme and the center keys the vowels. Word boundaries are not encoded. The output from these machines is a roll of paper on which the coded speech is printed. An example output from a Palantype Machine is shown in Figure 2.5(b).

Palantype follows relatively strict phonetic rules, but Stenograph uses more complex, less phonetic coding. Grand Jean, being weak on final consonants, is not appropriate for English. These machines provide a record of verbatim speech in the form of printed strips of paper that require significant skills to read. They are translated into orthography by trained operators. Automatic translation would clearly be valuable and was being investigated in the UK and the U.S. A major challenge with transcription of machine shorthand is to determine word boundaries, and this requires considerable (in the 1970s) computing power, and large amounts of storage for dictionaries. Research into automatic transcription of Palantype at the National Physical Laboratories (NPL) in the UK [Price, W., 1971] had not been taken up commercially due mainly to technological constraints, and to a (misguided) belief, prevalent in the UK, that tape recording would be cheaper and more effective [HMSO, 1977]. In the U.S., there was a much greater pool of Stenotypists, and this made Stenograph transcription a more commercially attractive proposition. C.A.T. (Computer Aided Transcription) systems, based on large and expensive (often time-shared) mini or mainframe computer systems which could not operate in real time, were beginning to be available [National Shorthand Reporter, 1974].

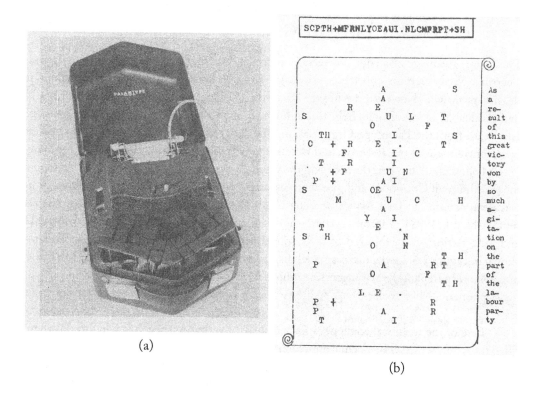

Figure 2.5: (a) The original Palantype Shorthand Machine; (b) paper output from a Palantype Machine.

Changing context can turn a failure into a success.

Neither the U.S. or the UK systems were appropriate for the situation I was investigating. In contrast to commercial use of machine shorthand transcription, the requirements for an aid for the deaf were a portable system that produced a readable—not necessarily correct—output in real time. I hypothesized that converting the phonemic codes to a readable form within the keyboarded syllabic structure could produce a readable output. The UK Palantype system uses a purer form of coding for the phonetic representation than Stenograph, Palantypists use very few abbreviations, and the operator training encourages more operator standardization than does the Stenograph system. Thus, a code conversion approach was feasible for Palantype, but any Stenograph transcription system was likely to require complex software and large dictionaries, and thus—in those days—a large computer system [Newell and Downton, 1979c].

2.4 DEVELOPING A FIRST PROTOTYPE WITH NO EXTERNAL FUNDING

The major advantage of a code conversion approach was that it could be done by relatively simple electronic circuits within a portable system. Joe King, an excellent undergraduate student, produced the first prototype [Newell and King, 1977b] with assistance and loan of equipment from NPL. We had invaluable and enormous help throughout all our Palantype transcription projects from Miss Isla Beard from the Palantype Organisation. She acted as an expert consultant and demonstration operator throughout our research, and was also Jack Ashley's personal Palantypist for many years.

Following demonstrations of King's system, we obtained a commission to develop a system for the House of Commons and also won research grants to develop the ideas further. A prototype was demonstrated to Jack Ashley and the Chief Whip, and the House agreed to purchase a system [Ashley, J., 1992]. A second system was designed and built by my colleagues, Andrew Downton and John Arnott, and subjected to a six-month trial in the House of Commons. This prototype, shown in Figure 2.6, used a plasma panel display mounted in a specially designed brief case. As can be seen in Figure 2.6(c), the output from this device is a simple code conversion and is syllabic and quasi-phonetic. Nevertheless, Jack Ashley was able to read this style of text after only a few hours training.

One of the technical challenges for a display of verbatim speech is what to do when the text filled the whole screen. The normal approach would be to move all the text up one line and write new data into the bottom line. This sudden jerky change, however, can disorientate the reader. Smooth scrolling was a possibility, but the speed of motion would be variable, and commercially available display systems did not offer such a facility. In addition, if the text moved up vertically, a reader who looked away from the screen could find it difficult to return to where they were reading. Leaving the text on the screen and writing over it from the top also proved confusing in practice. It was thus decided to modify the display by providing a "moving blank" of two lines situated immediately in front of any new data. This, together with a cursor, gave an unambiguous and clear indication of how to read the display at any moment in time.

2.5 NON-TECHNOLOGICAL CHALLENGES TO IMPLEMENTATION

The use of such a system within the Chamber of the House of Commons presented many political challenges. Objections raised included that:

- "Ashley would be at an advantage, therefore all MPs should have one",

- "He would have to have a seat assigned to him which was against the rules of the House" (Although woe betide any new member who took an established member's favorite seat. I also found that seats could be booked by inserting a card in them, before "prayers"),

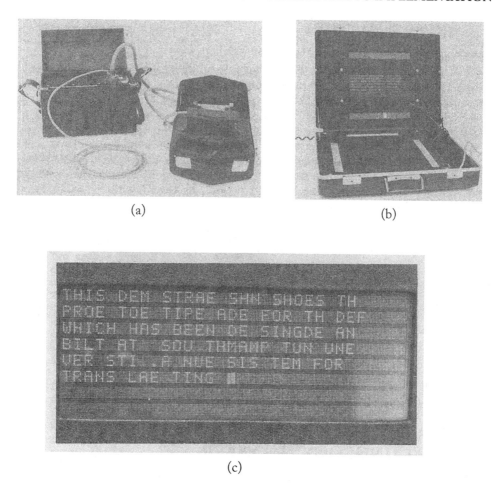

Figure 2.6: (a) & (b) First version of Palantype Transcription System; (c) output screen from Transcription System.

- "There would need modifications to the oak bench in the foreign press gallery" (where the Palantypist was to be situated).

As a non-MP, I had to obtain special permission from the Sergeant at Arms himself to sit on one of the "green benches" to try the system out, even though the House was not sitting at the time. At the Press Conference to launch the trial, it was commented that this was an historic day as "it was the first time in history that a member had had a specific seat assigned to him!".

We finally overcame all the objections and following a training period of approximately 20 hours, Ashley was able to follow all but the fastest speakers. The trial was a success and the service, with gradually improved systems, was continued for all Jack's subsequent career as an MP, both in the Chamber, in Committee and at one-to-one meetings. Ashley [1992] claimed that "It was a turning point in my life as an MP". A later system, shown in Figure 2.7(a), had a microprocessor and 20 kilobytes of storage [Newell and Downton, 1979c]. The output of this machine, shown in Figure 2.7(b) (which includes the effects of operator keying errors), was adequate for deaf people, but the commercial court reporting field had to wait until portable technology could support systems with large dictionaries.

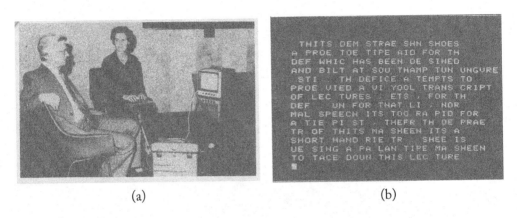

(a)   (b)

Figure 2.7: (a) The Rt. Hon. Jack (now Lord) Ashley using a computerized Palantype Transcription System; (b) the output screen of the Transcription System.

2.6   TECHNOLOGY TRANSFER

POSSUM Controls were licensed to produce the systems. The transfer of this technology was substantially assisted by Colin Brookes, a research student/assistant at the University, transferring to POSSUM Controls to manage their developments. The commercialization provided many challenges, as machine shorthand was not popular in the UK. Palantype machines had not been produced for many years, and there were no training schemes. Thus, not only did POSSUM have to re-design the Palantype machine itself, but also had to develop and market training courses for Palantypists. In the U.S., machine shorthand is very popular and thus all that was required was to develop electrical output for machines and transcription software. POSSUM systems were used in the UK in a variety of situations including by a deaf business man, many conferences, a telephone translation service for deaf people, and for live TV subtitling. The transcription software was improved and became adequate for commercial requires a high-quality output, because, if the recognition rate is less than 95%, it takes less time to re-type the script than to edit it.

Even with this improved system, POSSUM found it difficult to break into the Court Market. The Lord Chancellor's Office (who is in charge of Court Reporting in the UK) did not support this development. Officials believed that Tape Recordings and, eventually Automatic Speech Recognition, were the solution [Baker, 1966]. Tape recordings were introduced, but found to be more costly, less reliable, and not to capture important visual information (e.g., the witness pointed at the person in the dock). Their shortcomings were fully documented by Osmond's [1972] report of the Lord Chancellor's working party. POSSUM also found it difficult to persuade the Lord Chancellor's Office that a simultaneous transcript would be valuable (e.g., so that court officials could read it overnight). This was solved after a demonstration of the system to judges, who immediately saw the benefits. In later years speech recognition has been used in the U.S., but requires a trained speaker to re-speak the words uttered in Court.

> *Following fashion—even in research—may produce incremental advances.*
> *Swimming against the tide can lead to major advances.*

With funding from the National Research and Development Corporation (which became the British Technology Group), Palantype transcription systems were licensed to, and made commercially available by, POSSUM Controls Ltd. They also organized operator training. They offered transcription as a service to deaf people in a range of situations, including meetings, conferences, in a pilot telephone translation service and by a deaf business man [Hayward, G., 1979]. As technology improved, these systems incorporated very large dictionaries, and similar portable systems were developed in the U.S. for Stenograph machines.

Rapid reading of orthographic text requires good literacy, and thus can be difficult for pre-lingually profoundly deaf people, for whom a sign language translator is more effective, but both require a trained operator. The advantages of an orthographic output are:

- the words remain on the screen for a short time after they have been spoken, and thus the reader can briefly look away from the screen without missing words;

- orthographic output can be used for other purposes (e.g., retaining a record of the speech);

- a stenographer requires fewer and shorter breaks than a shorthand translator; and

- orthography can be used for "closed captioning" of television, whereas technology does not (yet) exist for transmitting sign language other than in the picture.

A French student at Southampton University showed that a Grandjean transcription system for the deaf was feasible [Sayi et al., 1981]. This was not taken forward, but CAT software (IBM-TASF) is now available which is compatible with the Grandjean shorthand machine. For his Ph.D. Colin Brooks [Brooks and Newell, 1985] investigated the potential of handwritten shorthand, but concluded that it was unlikely to be a viable alternative to machine shorthand.

In the U.S., Computer Aided Transcription (CAT) systems were developed for commercial applications, such as the law courts, and were not used for supporting deaf people until the technology could be made portable. In the UK the first commercially available CAT systems were small dictionary systems designed to support deaf people, and these became useful in traditional court reporting situations when large dictionaries could be included in such systems.

The major disadvantage with this research and development in the UK was that hand-written shorthand was much more popular than machine shorthand, and there was a shortage of operators: until POSSUM re-introduced it, there had been no training available for many years. In contrast Stenography is very popular in the U.S.—there are a number of companies producing training courses, machines, and CAT systems. Thus, as the availability of cheap portable computers which could host a full dictionary based CAT system became available, POSSUM's marketing edge in the field of supporting deaf people was removed and stenographic transcription, supported by a large technical base in the U.S., began to become available in the UK.

> *Support from potential customers is not a pre-requisite for successful research.*

At the time of applying for the grant, we did not have support from potential users (other than Jack Ashley), but the Research Councils at that time did not demand proof of commercial viability, and we were awarded a Science Research Council Grant "Simultaneous translation of machine shorthand for the deaf" (1977/79). With this funding we produced a prototype which worked in realistic environments, and also spent much time and effort in selling the idea to potential user groups, neither of which activities are "academic", and which tend to be squeezed out when University funding is reduced.

2.7   THE NEED FOR LUCK, FAITH, TIME, AND EFFORT

The Palantype Transcription story is the story of success being based on the results of what could be considered to be failures [Newell, A., 1988a], and had the following challenges:

- Although originally a great success and a very well designed system, Palantype Machine Shorthand never became very popular in the UK, and from the 1960's had begun to decline.

- The NPL work on CAT for Palantype had not led to a commercial product.

However, although automatic speech recognition systems only became commercially viable some 30 years after the research reported above, my knowledge of this area prompted me to investigate machine shorthand.

> *Success can be produced from commercial failures.*

The project required:

- a great deal of luck. A chance meeting with an MP and an excellent undergraduate student who produced the first prototype;

- a great deal of faith;

- financial support for untried ideas which had little support from potential users;

- time available for "academically" non-productive work including: marketing the idea, liaising with potential users, investigating companies and negotiating licenses;

- technical developments to produce a "pre-production" system that worked in real environments, rather than research leading to a proof of concept laboratory prototype; and

- acceptance of restrictions on publication due to commercial confidentiality.

It is interesting to speculate how difficult it would be to achieve this within the context of Universities in the 21st Century. Would there be time in an academics diary to do these essential, but academically un-productive aspects of a project of this nature? Would it be possible to obtain funding for an idea which had such a narrow focus, and no support (from the Lord Chancellor's Office) for the wider ramifications of the idea? An "Impact Statement" (as is now required by UK research councils) which reflected this reality would likely be seen as rather weak.

Another unanticipated spin-off from this research was a project supported by an Engineering and Physical Sciences Research Council into Automatic Speech Recognition. We used Palantype Machine Shorthand Transcription in a "wizard of oz" simulation of a "listening typewrite" to examine human factors aspects of a "listening typewriter" [Newell et al., 1991b].

2.8   COMMERCIAL AVAILABILITY

A range of portable communication aids for speech-impaired people are now available, including the Lightwriter, and are best seen at the commercial exhibitions associated with the biennial Conferences of the International Society of Augmentative and Alternative Communication. Machine shorthand transcription for hearing-impaired people is now routinely available in the UK using both Palantype and Stenograph systems.

CHAPTER 3

TV Subtitling for Hearing-Impaired People

A chance request led to an investigation on how to subtitle television for hearing-impaired people. This covered both the style of captioning that was most appropriate, and the development of efficient systems producing captions both for live and recorded programmes.

3.1 MOVING INTO THE FIELD

In the early 1970s, a lecturer in television engineering from a further education college approached me to ask if he could do an external Ph.D. Display technology had reached the state where it would be possible to create alphanumeric characters from ASKII code reasonably cheaply, so I suggested that he investigate the possibilities of "closed captions"—that is, subtitles which only appear on a set especially adapted/adjusted to display them. He rejected my suggestion of sending the caption data in the "lines above the picture", on the basis that these were used for engineering data, and investigated whether there were other parts of the signal that could be used.

Do not completely rely on the subject experts.

After about a year, the television authorities in the UK announced their Teletext system: a data service which utilized the area of the TV signal which I had suggested! Nevertheless, this "failure" began my interest in TV subtitles for hearing impaired people. The Teletext service had been designed to carry subtitles as well as data, and, from 1975, my group were funded by the Independent Broadcasting Authority to work with their "Oracle" Teletext team to investigate television subtitling.

The research questions that arise when considering subtitles for the hearing impaired are:

1. What text is appropriate for the subtitles?

2. How should the sutitles be displayed on the screen?

3. How can you create the subtitles efficiently?

4. How can you subtitle live programmes?

3.2   THE CONTENT AND COST OF TELEVISION SUBTITLES

The key questions are whether or not subtitles should be an exact copy of the dialogue, and whether exactly the same techniques could be used as those for foreign language films. The cost of preparing subtitles was also an issue at that time. With the technology of the day, it took 30 hours to subtitle a one-hour programme, at a cost that had been estimated at one third of one percent of the programme budget. On this basis, the broadcasters took the view that it was too costly to provide subtitles on all but specialized programmes [Newell, A., 1979a, 1982]. Also, with speaking rates (140 to over 200 words per minute) being much faster than typing rates (of the order of 60 wpm), verbatim subtitling of live programmes would be impossible. Even "ergonomic" keyboards cannot be operated at verbatim speeds [Newell and Hutt, 1979d]. At Clark School for the deaf in the U.S., a typewriter keyboard was being used to subtitle live programmes, but only the briefest of synopses was given.

There was some experience of subtitling for deaf people. TV guidelines had been produced in the U.S. but these were based on the needs of pre-lingually deaf children [Shulman, J., 1979]. In the UK, the British Broadcasting Corporation had significant experience with their subtitled Sunday evening "News Review" programme, and other programmes such as "Life on Earth". More research was needed, and the Independent Broadcasting Authority (IBA) commissioned my group to research the questions listed in Section 3.1. To this end we employed Rob Baker—a psycholinguist with experience of working with deaf people. This was my first experience of an interdisciplinary project, and having a subject expert as a full time member of the team proved to be crucial to the research.

The range of linguistic abilities of the viewers of subtitles for the deaf is much greater than for foreign language films, and thus we believed that, in contrast to this type of subtitling, the text would need to be an edited version of the sound track. Our early tests had shown that verbatim subtitles could put too high a reading load on pre-lingually deaf people who tend to have poor literacy. (The average reading age of a pre-lingually profoundly deaf school leaver had been established as approximately eight years, and, for most sign language users, English is a second language). The hearing viewer of foreign language films also receives a great deal of information from the sound track (including, knowledge of who is speaking, emotional content of speech, noises off and background music).

Editing, however, is not easy—language does not have an homogenous structure, its rules changing dramatically with the talker, the assumed listener, the message and the environment. Thus, in certain circumstances, a single word can make an important difference—the word 'not' obviously changes a sentence, but some adjectives also have great importance (e.g. "the Prime Minister said that there was (little) (real) truth in the statement....."). It is also clear that the soundtrack has different roles in different programmes. In sports, it is mainly there to communicate excitement, and a most bizarre form of English can be found in football commentaries, which often have very low information rates, and strange syntax. Music also makes an important contribution to mood. The question arises of how such information should be transmitted in sub-titles.

We experimented with showing hearing impaired audiences verbatim subtitles, and compared these with various edited subtitles and with ones where a commentary rather than a version of the words spoken was provided. The commentaries were not liked, and the edited subtitles were found to be easier to read. The mismatch between lips and words on screen were not found to be a significant problem, and we found a general preference for edited subtitles, but only edited to produce a manageable reading rate [Baker et al., 1981].

Different views were expressed for different programme types, e.g., viewers preferred edited subtitles for news and serious drama, whereas verbatim subtitles were preferred for chat shows and comedy. A further interesting finding was the difference between the impact of the spoken and written word, the most obvious example being the use of swear words. These have a greater impact when read as subtitles rather than when heard. There were also cases where the subtitler had to decide whether to retain the words used with the danger of changing the impact, or retain the impact and change the words [Baker and Newell, 1980]. On the basis of these results Rob Baker [1981] produced a comprehensive set of guidelines for the IBA.

In the UK there is a tendency to provide short, uncomplicated sentences, but in the U.S., fully verbatim subtitling is preferred. This reflects a different compromise between making the subtitles easy to read and being "faithful" to the original spoken words.

Reading rapidly changing subtitles and watching the picture is not a trivial task, and viewers need all the help they can get. It is very difficult to watch the picture and read rolling subtitles, and our experiments showed that splitting the subtitles into meaningful units was helpful. There is also a question, particularly with live subtitling, whether the subtitles should follow the speech or be presented as meaningful units. We also found that a rectangular box with a black or misted background closely fitting round the subtitles increased readability. The changing shape of this box had the advantage that it gave a clue that the subtitle had changed. On the other hand, retaining a subtitle over a video cut had the effect of the viewer reading the subtitle again.

3.3   LIVE SUBTITLING

In the 1970s a number of groups were investigating subtitling live programs. At Leicester Polytechnic, Booth and Barnden [1979] worked with the BBC. They used a Palantype speech transcription system they developed, which ran on a large computer with an 80,000 word dictionary, and inserted single line subtitles into the television signal. Independent Television used the Palantype Transcription system that we had developed for Jack Ashley. This had a microprocessor with a 1000 word dictionary plus transliteration software and produced multiline subtitles. In 1979, the U.S. government set up a National Captioning Institute, with staff of 40, captioning 16 hours of programmes per week [McCoy and Shumway, 1979]. They produced a set of guidelines in 1980, and were investigating live subtiting using the American Stenograph system. Somewhat later, the Dutch investigated the use of the Velotype (1983) keyboard for live subtitling.

All live subtitling systems have the disadvantage that the subtitles are a few seconds behind the words spoken (this could be solved by delaying the program by a few seconds, but this is unlikely to be acceptable to the broadcasters). The potential mismatch between the picture and the subtitle can be confusing, and potentially embarrassing. ORACLE (Independent Television's telextext system) was used to subtitle the 1981 Royal Wedding, using a QWERTY keyboard and the NEWFOR systems (see below). A significant part of the proceedings could be prepared in advance, but not all. A video clip of a county home of Lord Montbatten was shown, with the commentary that "this is where the Royal couple will spend the next three days". Unfortunately, this caption appeared above the next video clip that was of a four poster bed.

3.4   A SUBTITLE PREPARATION SYSTEM

The Southampton group did not think that the current systems for preparing subtitles for television were particularly efficient, and suggested a research programme to develop a specially designed system. The broadcasters, however, were not convinced of the value of such work. Thus, rather than a funded research programme, this research was begun by a research student. It started with a detailed study of the tasks involved in subtitle preparation. These were:

- programme preview;

- text input and formatting;

- synchronization with video/soundtrack, and timing; and

- review and modification where necessary.

  Lambourne et al. [1982a] listed the important factors to take into account:

1. the programme style (e.g., drama, documentary, etc.);

2. speech rates and dynamics;

3. syntactic complexity;

4. dialect and idiom;

5. vocabulary complexity;

6. position on screen;

7. kind of background to provide;

8. how many words/subtitle;

9. how far can speech be edited without loss of information;

10. do sports commentaries require a different approach; and

11. how to deal with offstage voices and noises.

Examining the different ways subtitles were produced led to Lambourne developing NEW-FOR (New FORmatter—marketed by VG Electronics, Hastings, Sussex). This was an input console serving as a front end to existing subtitling systems. It provided automatic formatting (using syntax analysis to highlight significant markers such as phrase, clause, or sentence ends), and the use of geometric shape analysis for line division, insertion of colour (to denote speakers), positioning of subtitles on the screen, and calculations of on-air display time. Abbreviation expansion was also offered. After being shown preliminary results of this research, the Broadcasters funded further development. NEWFOR reduced the time to produce subtitles from 27.5 hours/per programme hour to 15 hours.

The most important aspects of this research were the multi-disciplinary approach of detailed research into the requirements of deaf viewers and the analysis of the captioning process. We believed that these had not previously been done, and they led to an effective and efficient system. Some 30 years later, the original research student Andrew Lambourne continues to market subtitling systems. This provides yet another example both of the potential client not supporting the work in its early stages, and the advantage of researchers moving from academia to industry to facilitate the technology transfer.

In the 1970s, broadcasters had said they could not afford to subtitle programmes, but by 2008 100% of B.B.C's output was subtitled (although this was not the case in some other UK broadcasters' output). The importance of subtitling is underlined by the BBC having more complaints about the audibility of speech than anything else. Particularly for older people, it is often a combination of hearing impairment and the cognitive capacity to understand heavy accents and/or poor diction, which produces the problem. In this case, subtitles can provide a more effective solution than "clear audio" or reducing other sounds in the audio output. Subtitled television has also been found to be very popular in public houses and other noisy environments. This is another example of a system designed for a disabled group providing a service for non-disabled people (see Chapter 10).

CHAPTER 4

Word Prediction for Non-Speaking People and Systems for those with Dyslexia

A major design challenge for the developers of communication system for non-speaking people was to increase the rate at which non-speaking people could "talk" via these systems. PAL, the predictive adaptive lexicon, showed how this could be achieved by word prediction. This system was found to provide a writing aid for people with dyslexia, and led to the development of other software for this group of people. An alternative word prediction system was reproduced many years later as a ubiquitous mobile phone application.

4.1 SUPPORT FOR NON-SPEAKING PEOPLE

Augmentative and Alternative Communication (AAC) devices are designed to assist people who cannot speak due to physical and/or cognitive/language dysfunction. The low-tech version of an AAC device, a word board, consists of a board with letters of the alphabet, words and/or pictures to which the non-speaking person points. Early research and development focussed on systems which were essentially an electronic replacement for a word board [Copeland, K., 1974]. These had either a printed or visual output, but more modern employ computers and speech synthesis. They can be controlled by a normal computer keyboard, but specially designed large keyboards, and coding systems such as Morse code are available for physically disabled users, and a single or two keys can be used to access the cells in a scanned matrix [Korba et al., 1985]. The impact of these systems was enormous. They changed lives dramatically and provided an example of the importance of spoken communication to the perceptions people have of their fellow human beings [Beukelman and Mirenda, 1992].

> *"However severe your physical disabilities are, it is your speech that will disable you the most".*
> **(Geoff Busby, B.C.S., private communication)**

The advantages of AAC devices having synthetic speech output is that this is much closer to natural speech communication than orthography. The overriding difference between natural speech and AAC produced speech, however, is speed. Natural speech is usually in the range 150—200 words per minute whereas even a good typist can only type at 60+ words per minute, and a physically disabled AAC user may only be able to hit less than 1 key per second. It takes patience and skills to communicate with someone "talking" at 10 words per minute, and such users can have great difficulties in communicating their thoughts to all but the most sympathetic partners. One of the most well known AAC users is Professor Stephen Hawking of Cambridge University—but he is a very atypical user. He has excellent literacy skills, and, as he is very well known, people are prepared to wait a long time for his words of wisdom. Kraat, A. [1985] reviewed the state of the art in AAC, and it was clear that, although it would be straightforward to connect a speech synthesizer to a computer system to allow a person to "talk", such a system would not provide a very satisfactory prosthesis.

In order to design a successful communication aid, it is first necessary to examine the communication act. Light, J. [1988] divided communication into:

- communication of needs and wants;

- information transfer;

- maintaining social closeness; and

- performing social etiquette.

The designer of an AAC device also has to bear in mind other types of information contained in natural speech. Speech communication contains a great deal more than the words which are spoken [Morris, D., 1982]. It is also used to convey other information, which can include:

- the personality of the speaker;

- the emotional state which the speaker wishes to communicate;

- the geographical, ethnic and socio-economic background, and often the identity of the speaker; and

- the relative status between speaker and listener.

An additional factor that has to be taken into account is "for whom the device is primarily intended". Communication is between at least two people and thus "users" of the aid includes

- the "primary user"—the disabled person;

- friends and family;

- therapists;

- medical staff; and

- strangers.

These groups may have very different requirements. The situation is made more complex by the fact that the decision makers and the purchasers of the aid are more likely to be therapists or medical staff than the primary user. This can produce conflicts of interest; some clinicians can prefer a patient to be non-communicating, and there have even been reports of "non-accidental damage of aids" [Newell, A., 1984]. McGaffey et al. [1991] commented that "...clinicians don't really want us to communicate they just say they do".

4.2   DEVELOPING AAC DEVICES

The design of an effective AAC device is not trivial and requires input both from clinicians and technology development teams. Such collaborations have been reported in the major journal in the field—the International Journal of Augmentative and Alternative Communication. Early work by clinicians included important papers by Light, J. [1988], Kraat, A. [1985], and Alm and Parnes [1995]. Significant technical advances were made by teams led by Greg Vanderheiden at the Trace Centre (University of Wisconsin—Maddison), Tijus Soede at the Instituut voor Revalidatie-Vraagstukken at Hoensbroek in the Netherlands, engineers at the Hugh MacMillan Medical Centre in Toronto, Higginbotham at University of Buffalo, and Rick Foulds as well as the team at Dundee University.

Research priorities for the field were laid down at the Visions Conference [Mineo, B., 1990] organized by Rick Foulds in University of Delaware March 1990. I outlined my views at this conference and in the Jansson Memorial Lecture to the (UK) Royal College of Speech and Language Therapists later that year [Newell, A., 1991]. A major research priority was to reduce the keystrokes necessary to produce speech output and/or increase the speed with which users could output synthetic speech.

4.3   WORD PREDICTION AND ASSOCIATED TECHNIQUES

As soon as it became technically possible, researchers investigated methods of increasing the speed of AAC devices. Spoken and written language is a very redundant coding method and this makes it possible to reduce the number of keys needing to be pressed to create words. Baker [1982] developed an iconic coding system, which used a sequence of icons to access words and phrases. This was very successfully marked under the name of "Minspeak" by the Prentke-Romich Company, but did involve a significant amount of training to learn how to remember these sequences.

Other researchers investigated the use of alphanumeric coding schemes. These included using N gram techniques to predict the next letter of a work being typed. Heckathorne and Childress [1983] used trigrams to insert predicted letters into very accessible cells in a scanned matrix. Short-forms and abbreviations were also used to reduce the number of keystrokes needed. Vanderheiden [1984], and Demasco and McCoy [1992] used a technique which they called "compansion" which allowed the user to input words in a reduced form, and thus generated messages from keywords or telegraphic input words [Cushler et al., 1996].

Although they reduced the number of keystrokes required, the problem with abbreviations is that they have to be learnt, are only appropriate for frequently used words, and thus only give a limited advantage with large vocabularies. Another way of reducing keystrokes is word prediction and a number of researchers investigated this approach [Colby et al., 1982, Newell et al., 1995, Vanderheiden and Kelso, 1987, Whitten et al., 1982]. These techniques were applied both to keyboard operated devices and to devices with a scanned input. The user is presented with a menu of words predicted based upon the letters which have been typed, the list being updated as each new letter is typed.

Early predicted systems had hand built dictionaries, but, following a suggestion from a psychologist, Adrian Pickering and colleagues in Dundee produced an adaptive prediction system called PAL (Predictive Adaptive Lexicon) which had an in-built dictionary, but also captured new words and inserted them into this dictionary. The major advantages of PAL were that (a) the frequency of word usage was updated as the system was being used—and thus reflected the individual users vocabulary and (b) predictions were based on how recently the word had been used. (We noticed that, in written text, if an unusual word is used it is often repeated later in the text.) This ensured that the predictions were closely matched to the words that were being used at any one time as well as the vocabulary patterns of a particular user [Newell, A., 1987a, Swiffin et al., 1985].

We needed to know how many predictions we should offer to the user at any one stage. The greater the number offered, the fewer keystrokes would be needed to predict a word, but larger lists would increase the time taken by the user to check whether the word was in the list. We investigated the effect of offering different numbers of predictions, and noted that the key-saving curve had a knee at approximately five words. We decided that this was an appropriate compromise. The effectiveness of such a system is critically dependent on the visual interface, and the time needed to scan prediction lists depends on the orientation and position on the screen of predictions. PAL, shown in Figure 4.1, had a vertical list of five words (which can be perceived in a single glance [Miller, G., 1956]), with the word length providing a very obvious visual cue. They were offered immediately below the cursor so reducing the need to change the user's point of gaze. In contrast, some contemporaneous systems offered their predictions in a horizontal list along the bottom of the screen. Higgingbotham, D. [1990] studied how different device characteristics (prediction lists, keyboard layout) impact on an individual's ability to understand and produce messages, and Light et al. [1990] discussed the costs of the associated learning required. Clearly, good screen design makes a great deal of difference in these systems.

These data provided an early indication that detailed screen design is vital for an efficient system. We determined that a key-savings of between 29 and 41% could be obtained by using PAL. (This can be compared to a "perfect" word prediction system which would require one keystroke per word giving an average key-saving of 82%).

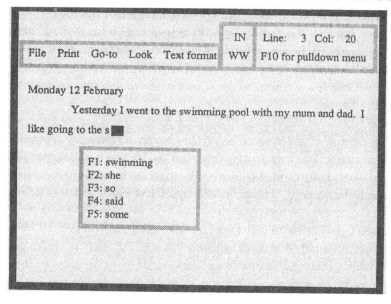

Figure 4.1: The original PAL word prediction system.

4.4 PREDICTION SUPPORTING DYSLEXICS

We, and other researchers designing predictive word processors, were focussed on reducing keystrokes and improving speed for physically disabled children and adults. Our research team, however, included a special education teacher, Lynda Booth. Her role was to take our prototypes out to schools to investigate how they worked in practical situations. Lynda took PAL to a special school to try it out. After a few sessions, it became obvious to her that physical disability was not the only issue and many children displayed classic learning difficulties. Lynda noticed that children with spelling disorders and language dysfunction were looking to the predictions for assistance.

> *Classic discovery—to search for what you know (say a new way to reach India) and find something you did not know was there (America).*
> **The Black Swan, N.N. Taleb, Penguin books, 2008, p. 166**

PAL was thus trialed with a wider group of users including those with spelling difficulties. Improvements of up to 65% fewer errors were found. An example of the effect of using PAL on a cerebral palsied non-speaking child is shown below.

Without Prediction:

today it souning. It is deebb. I may go home eithl. I like son. It is coil and whiith

With Prediction (which was typed at twice the speed of the above):

Today it is snowing. It is deep. I may go home early. I like snow. It is cold and white

A 15-year old with specific learning difficulties produced the work shown below

Without Prediction:

My farm is arbo it is ol plelbo and can be yoot for seros it as a qora duon ner the bottom

With Prediction:

my farm is arbo it ol ploughable and can be ytt for sero it at a quarry down need the bottom.

A more extreme example was a 10-year old boy with severe developmental delay with whom several reading schemes had been tried without success. He was introduced to PAL with a dictionary of 10 personal words with a wide range of word length, so he could easily differentiate them. After six months he had a working written vocabulary of 150 words and was writing simple sentences using PAL.

The use of prediction to assist spelling had not been foreseen by developers. Traditional spellcheckers are little help for those with bizarre spelling, such as dyslexics, but some dyslexic children and adults can recognize correct spellings even if they cannot create them [Ellis and Young, 1988]. Many are better with the beginnings of words than the middle and ends. We conducted formal trials of the use of prediction with poor spellers [Newell et al., 1992] and found that Lynda's vision had opened up a whole new use of predictive text systems. In a later evaluation study, nine case studies were reported in which PAL not only greatly improved the speed at which children with poor keyboarding skills could produce written work, but also significant improvements in the quality of this work were seen particularly for those children with spelling difficulties [Newell, 1991]. A full analysis of the results of these case studies can be found in Newell et al. [1991b].

> *Learning to read by writing.*

We also found some evidence of an increase in vocabulary, and improvements in unaided written work, which led to children developing more positive attitudes towards developing their own literacy, and PAL was also found to help "reluctant writers". Thus, bizarrely, the use of PAL has enabled children to learn to read by writing! PAL was licensed to Lander Software, Glasgow Scotland.

4.5   EVALUATION OF THE EFFECTIVENESS OF PREDICTION

There were some disputes within the research field during this period. Keystroke saving was seen as the major advantage of prediction and some researchers had used simulations to determine the advantages of various coding schemes. It was suggested that word prediction was less effective than coding schemes such as Minspeak [Baker, 1982], because of the time to scan lists. This was questioned by Newell et al. [1992] on a number of accounts. Issues included that it did not consider the effect of word prediction on improving written literacy, and writing in general (which code driven systems were less likely to do), that word prediction requires less cognitive effort, and also remembering which icon corresponds to the word required may take time. McNaughton, K. [1980] also pointed out the dangers of short-term evaluations and the importance of long-term evaluations, not only to allow adaptive systems to fully reflect the linguistic behavior of the user but also allowing the user to become fully familiar with the system.

> *Success depends on what is measured.*

This was confirmed by one of our researchers—Annalu Waller. Cerebral palsied herself, she joined the Dundee team in 1989. She had been using abbreviation expansion, but began to use PAL on a daily basis. Her first reaction was that the time to scan the lists meant that she was typing slower, but her typing rate was actually increased and she noticed a reduction in fatigue. She found prediction much more useful than abbreviation expansion, and could not envisage being able to learn the codes for thousands of words. She eventually used a system which had three dictionaries—professional correspondence (5560 words), research (6205 words), and personal correspondence (6646 words).

This experience underlined the fact that even such apparently objective measures such as key-saving and rate of composition need to be verified by long-term case studies. Having a disabled researcher in the team, who was both being a real user and having a detailed knowledge of what was technically feasible, provided an ideal environment for the latter stages of the development cycle. It is not easy to bring such a team together but the rewards are significant [Waller et al., 1991a].

Swiffin et al. [1987] and others experimented with improving prediction by adding syntax rules to predictive systems. Guenther's [1993] KOMBE word prediction system required the user to input linguistic information—e.g., morphological and semantic information—but this requires either the user or the carer to have a significant degree of linguistic training. Others included VanDyke et al. [1992] and Hunnicutt, S. [1989]. Morris et al. [1991] introduced a simplified syntax model into PAL. This proved to be more pleasing to use, as really inappropriate words were not predicted. It reduced keystrokes by approximately 5%, and with some users reduced their syntax errors, (e.g., reduced the omissions of function words). Adding syntax rules did not increase speed dramatically, but may have assisted the users in their English composition.

This particular investigation was interesting in a generic sense. We, and others such as McCoy and Demasco [1985] and Copestake et al. [1997], had investigated the use of Natural Language Processing techniques in AAC systems [Newell et al., 1998]. For example, we used incomplete syntax models and a subset of word classes in our predictive system. This meant that some of the predictions were incorrect, but this only marginally reduced the effectiveness of the system. Such models would have been inadequate in any linguistic forum as they were not applicable to the whole domain of spoken language, but they were effective within our context. In this sense this was another example of the field taking advantage of a "failure" in another field.

Wright and Newell [1991] developed a spell checker which used phonetic as well as lexical matches for people with severe dyslexia: /eggriclchr/ for agriculture, /fysis/ as physics. This met with a favorable response from an evaluation with a group of 6 children. In a pilot evaluation, correct spellings were predicted 84% of the time. This system was integrated with PAL on an experimental basis, but it was not commercially exploited.

4.6   OTHER TECHNIQUES TO SUPPORT DYSLEXICS

The research reported in Section 4.4 caused us to consider other software applications that could assist people with dyslexia. Peter Gregor (who had done some important research into computer based interviewing particularly for people with emotional and behavioral problems) and Anna Dickinson, examined how their reading experiences could be facilitated. It was known that changing the visual aspects of text could assist some dyslexics, but there were no clear guidelines as to what characteristics of the visual scene were most helpful, and, in any case, these varied from individual to individual.

The approach adopted was to implement Shneiderman's theory of 'Direct Manipulation'. Shneiderman [1998] describes direct manipulation as: "rapid incremental reversible operations whose effect on the object of interest is visible immediately". "SeeWord" was a text reader that allowed users to easily configure the appearance of text by using on-screen objects which, when manipulated by the user, immediately affect the appearance of the document: this included fonts, colors of text and background, spacing between characters, words, lines, and paragraphs [Gregor et al., 2003]. Figure 4.2 shows examples of the control screens of SeeWord. The user was able to select, by experimentation, the settings that best suit them. These settings were saved and made available

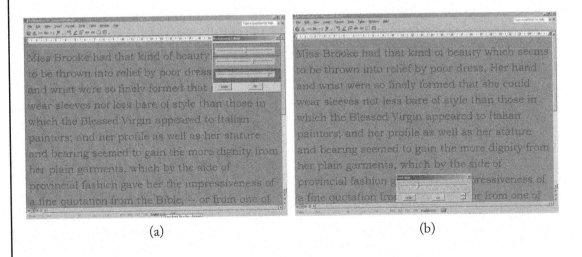

(a)   (b)

Figure 4.2: (a) SeeWord—sliders to change font and background colours; (b) SeeWord—a slider to change the font size.

each time that person uses the word processor. The system was evaluated with 12 dyslexic computer literate students in higher education using think-aloud protocols, questionnaires and interviews. A second prototype, developed as an add-on to Word for Windows, provided a facility to enhance characters prone to reversals (e.g., b, d), by using color font type and size. This was evaluated by seven dyslexic users with an age range of 15 to 30 years, and provided unanticipated results—the users reported the value of this approach was that the sporadic coloring broke the text up and resulted in them being less likely to get lost—that is the system was reducing fixation problems rather than recognition problems. There have been more recent claims that some dyslexics find reading easier on an iPod, presumably because less text is displayed, and there are less visual distractions than on a conventional computer screen.

SeeWord was found to have potential to make computer-based text significantly easier to read than printed text, as well as improving the usability of computer word-processing systems for a range of dyslexics. Importantly, this research highlighted the individual nature of the disorder, and the diverse characteristics of any effective user interface. It illustrated the importance of designing for a diverse user base, and how a standard user-centered design methodology is not always appropriate for such user groups.

4.7 DISAMBIGUATION

In the early 1990s, an alternative approach to keystroke reduction was developed for those people with poor hand control who could only use a keyboard with a small number of keys. Arnott and Javed [1990, 1992b] from the Dundee Group, and a small number of other research groups in the world investigated disambiguation as a way of improving typing by people [Foulds et al., 1987, Kamphuis and Soede, 1989, Levine et al., 1987, Minniman, S., 1985]. In these systems each key is labeled with a number of alphabetic characters (as occurs, for example, on standard telephone keypads). The user thus enters an ambiguously coded word, and the system disambiguates this to produce correct English words.

These researchers investigated the use of both word and N-gram character disambiguation with a range of number of keys. Foulds et al. [1987] showed that 92–95% of English words have a unique N-key code on a 12-key keyboard, and this proved the most successful manifestation of disambiguation. Arnott et al. evaluated their system with four cerebral palsied users [Arnott et al., 1992a], who found it to be an effective and efficient way of producing text.

Technology to support disability can lead to innovations in mainstream products.

Some years after this research had been published, the idea was taken up by the Tegic company and their T9 appears in most mobile phones to assist text prediction. This has become one of the most widely available examples of Assistive Technology, and shows how disabling a telephone keypad is for inputting text. Disambiguation techniques have also been used to improve typing performance on touch sensitive screen keyboards. ShapeWriter disambiguates gestures made across the keyboard, and BlindType uses disambiguation to provide a text input "app" that is very tolerance of keying errors.

It is interesting to note that text telephony was used by deaf people for many years before text messaging by mobile phone became popular. Newell [1993a] suggested disambiguation as "valuable in any situation in which there is insufficient space for a full alphanumeric keyboard"—but did not envisage the popularity of it in mobile phones in the future! In a discussion of the use of broadband in an assistive technology environment, McKinlay et al. [1995] suggested prediction in text telephones for deaf people and computer based text messaging by disabled people long before the advent of the popular text facility on mobile telephones. Despite deaf people having found text messaging very useful, however, engineers did not think to build this facility into mobile phones (it was a by-product of an entirely different technical requirement of such devices). The McKinlay et al. [1995] paper also suggested collaborative use of prediction (i.e., the communication partner choosing predictions) and the advantages of including video information in a broadband systems. Such facilities are now available via Skype.

4.8   BEYOND WORD PREDICTION

Word Prediction provided speed increase in the output from AAC devices, a reduction in keystrokes needed and user fatigue, and also provided assistance for those with spelling difficulties. The maximum keystroke reduction achievable using these methods, however, is not sufficient to bridge the gap between typing speeds and normal speaking rates of 180 plus words per minute. A physically disabled person may type less than 10 words per minute and a good typist can only achieve 60 words per minute. Different techniques were thus required to tackle this challenge.

CHAPTER 5

Providing Reusable Conversation for Non-Speaking People

Only modest increases in speed of communication are possible using word prediction. Examining the real needs and wants of non-speaking people, together with the academic fields of conversational analysis and speech act theory, led to the development of "CHAT" (Conversation Helped by Automatic Talk). This system used a model of conversation to provide effective and efficient access to pre-stored speech acts. It was based on the concept that the pragmatics of conversation are often as important as the semantics. The ideas it embodied were not without criticism in the field.

5.1 AN APPLICATION OF CONVERSATIONAL ANALYSIS

Very slow speech requires a great deal of patience from the conversational partner and, particularly in group settings, AAC users find it very difficult to keep up with conversations with natural speakers. There is a tendency for natural speakers to break into conversations whilst an AAC user is creating an utterance. Word prediction did give advantages in terms of keystroke reduction, but prediction of phrases and sentences could significantly increase utterance speed. We thus examined natural conversation with such a concept could be part of AAC devices.

John Arnott was browsing in the library and came across the discipline of conversational analysis, and Alm, N. [1985] commented that "a consideration of the characteristics of natural communication paid great dividends in understanding the needs of communication aids". Our AAC research was also greatly assisted by Bernadette Brophy-Arnott, a research speech therapist in the team, who participated in many of the research projects. She also conducted a very detailed survey of the local communication impaired population that provided valuable information to the team [Brophy-Arnott et al., 1992].

As part of his postgraduate research, Alm, N. [1987] conducted a literature review of conversational/discourse analysis to ascertain whether the concepts within conversational analysis would be useful in AAC, and initiated a numbers of field studies. He also worked with psychologists Todman et al. [1995] and Portia File, a computer scientist, to investigate strategies for effective communication. They worked closely with a number of non-speakers in this work. In the U.S., Light et al. [1985] studied the way non-speakers negotiated their communication goals. Up to this time, communication devices had attempted to provide help for non-speakers at the level of producing individual words, but we believed that this method was based on a simplification of what actually happens in unaided speech communication.

Natural language research provided a number of important insights that proved important to this research. These include:

- an enormous amount of natural language is formulaic, automatic and rehearsed rather than propositional, creative and freely generated (Fillmore quoted in Gumperz, J. [1982]);

- a significant percentage of conversational language is highly routinized into pre-fabricated utterances [Stubbs, M., 1993];

- opening and closing sequences are important, but are frequently ritualistic encounters and formulaic in their context [Laver, J., 1974];

- non-assisted conversations include stories, gossip and jokes that are made up of a succession of sentences in a highly organized social activity [Clarke and Clarke, 1977];

- conversational narratives provide a communicative way of forming experience and relating past experiences [Quasthoff and Nikolaus, 1982] and play an important part in an individual's social and educational development; and

- At 120–200 wpm speech is too rapid to allow a speaker to consciously play the speaking of each word separately (claimed by Vygotsky, L. [1962]).

Very few AAC users in the 1980s and 1990s engaged in gossip, stories, or jokes because the effort needed to produce the text was too great. In addition to slow speech rate, AAC users' speech include long gaps whilst they are preparing to reply to a comment made by a speaking person. Unfortunately, in non-aided conversation, silence is perceived as undesirable, or even can be perceived to have a meaning [Newman, H., 1982]. At the very least, such silences can make the non-aided speaker feel uncomfortable, and encourage them to interrupt.

An AAC user will be seen to speak very slowly with longs gaps between utterances, and people often mistakenly attribute lack of intelligence or low social skill to AAC users on the basis of their poor communication. Speech is very important in the initial perception of personality, and there is a tendency to judge a person's intellect on the basis of the speaking style. This perception can be difficult to change. In some sub-cultures initial perception is reinforced by symbols—e.g., the rank badges on military and police uniforms, and for civilians expensive designer clothes and watches, but non-speaking people often present other non-verbal messages which can be negatively received.

*"If I were to be granted one wish and one wish only, I would not hesitate for an
instant to request that I be able to talk if only for one day or even one hour".*
**In: I raise my eyes to say yes—Sienkiewicz-Mercer R. & Kaplan S.
Houghton Mifflin, 1989.**

5.2   PERCEIVED COMMUNICATION COMPETENCE

McKinlay, A. [1991], a philosopher and conversational analyst, was appointed to the research team,
and he made the distinction between "communication competence" and "perceived communication
competence"(see also Zebrowitz, L. [1990]). This distinction can be seen in the film "Being There",
and is discussed in Kosinski, J. [1968]. Chance the Gardner, played by Peter Sellers, is a simple-
minded recluse, with a rich adopted father. He has had an upper-class social training but his whole
educational experience is drawn from watching television and cultivating a garden. He is raised to
power and success by a combination of mistaken identity and, crucially, an ability to use appropriate
conversational ploys. Chance had a high level of perceived competence, but a low level of actual
competence. In the film this mistaken perception carried him to great heights.

When Chance was asked by the President of the U.S. "what do you think of the bad season in
(Wall) Street", his reply, "as long as the roots are not severed all is well and will be well", was hailed
as a very important contribution to the debate on the U.S. economy. Once Chance had established
his reputation, all his answers are automatically assumed to be intelligent, and inappropriate remarks
misunderstood as metaphoric. It did not matter what Chance said as long as he said it fluently and
with authority. Many AAC users are in the opposite position to Chance, in that the perception
of their competence is much less than their actual competence. The reputation of the AAC user
Professor Stephen Hawking ensures that this seldom occurs in his case.

5.3   TECHNOLOGY TO ASSIST DECEIT

Providing a perception of communication competence is a further challenge for the designers of AAC
devices. In essence, in the language of theatre, we need to design devices that will suspend the disbelief
of the conversational partners of AAC users. We need to investigate the ways certain professionals
use technology to increase the perception of their communication competence. Lecturers use notes,
and newsreaders and high ranking politicians use almost invisible autocues. They also put in a lot
of practice for their apparently spontaneous speech; e.g., the UK Prime Minister rehearses "Prime
Minister's Questions" in a session where his assistants throw questions at him that they think will
be asked in the next session—in this way the P.M. can give a well-rehearsed answer to an apparently
unknown question. The same technique is used by stand-up comics to shape what appear to be
entirely improvised exchanges with the audience. Poets are skilled in putting a great deal of meaning

into a few words—a technique which would be extremely useful for AAC users. Perhaps, in addition to speech therapists, AAC uses should have lessons from script writers and poets.

> *"All the world's a stage, And all the men and women merely players".*
> **Shakespeare. As you like it, 1, vii.**

An early experiment where the disadvantages of being an AAC user was essentially removed was performed by Stephen von Tetzchner from the University of Oslo [Von Tetzchner and Grove, 2005]. They set up an email group of able-bodied and disabled users. Because the messages transmitted within the group did not give any clue to how long they had taken to compose, the disabled users essentially were judged to have the same communication competence as the able-bodied users and thus achieved equal status with them. This was very much appreciated by those participants. It is interesting to note that the same effect would occur with disabled participants in modern social networking sites such as Twitter.

5.4   "CHATTERING, NATTERING AND CHEEK"

Social bonds, or the lack of them, are a major factor in how society views non-speaking people, and the inability to make successful social bonds can be very damaging to a non-speaking person's self image [Newell, A., 1992a]. Social bonding includes "chattering, nattering and cheek", and to successful socially bond a person has to:

- make people talk;

- stop them talking; and

- have some effect on what they say when they are talking.

Important characteristics of conversational control and bonding include:

- saying the right things—including "stroking behaviour";

- obeying turn-taking rules; and

- rapid delivery and careful timing.

A person needs to communicate:

- their personality;

- their attitudes and feeling;

- their mood;

- their relationship with the current conversational partner;

- the role they have in society; as well as

- simple information and requests.

> *Relationships can be more important to people than information.*
> **(Hence the popularity of social networking sites such as "Facebook")**

Crucial questions are:

- "what is the aid going to be used for"? and

- "what messages need to be made available"?

There is a tendency to base message sets on a too superficial view of what is needed. Decisions about message sets require a serious consideration of the expectations of the user, and the possible communication environments to which they may be exposed. Spelling out messages gives maximum flexibility, but is very slow and tedious and may really only be needed for "important messages" rather than the day-to-day chit-chat which is a major part of most people's spoken discourse. (It is difficult to develop a relationship with someone if one is confined to traditional AAC messages about food, toileting and medical needs). Developing a message set needs a clear and empathetic view of what the user may want.

Any AAC device should contain appropriate words, including positive expressions of emotion and, if requested, swear words. (One early commercially available AAC device censored expletives in the hardware by substituting keyed-in swearwords by other words, such as 'fudge' or 'sugar'!). Also, carers may be less than enthusiastic about including non-socially acceptable phrases within an AAC device. Alm, N. [1993] includes an interesting story of a school teacher's refusal to agree to include a swear word within one of her pupil's devices. Pullin, G. [2009] reports of a high functioning AAC user who had a simple device designed and built for her which, when she hits it with her head, says "Sonia says SOD OFF". Users, however, do need to be taught the rules of conversation, and pragmatic skills, including when not to swear, as these are unlikely to be learned by simply observing conversations.

The obvious way of increasing the speech rate of AAC users and reducing the length of silences would be to enable users to store pre-prepared utterances, and access and transmit these phrases and sentences rather than encoding individual letters of words. The challenge of such an approach is that a large number of phrases need to be stored, and a significant memory load may be required to access these phrases. It would also be valuable if the AAC system were to encourage natural patterns that occur in speech communication, thus making systems more effective for promoting appropriate conversational behavior. These include appropriate turn taking, maintaining place in a conversation, initiating and encouraging conversations, and delivering specific messages, requests and responses, as well as encouraging the conversational partner to speak (such as "open" answers to questions, rather than "closed" remarks such as "yes" and "no").

In essence, this redefines the "rate problem" as "how do we enable an AAC user to quickly initiate appropriate socially bonding communication acts".

5.5   DIFFERENCES IN CONVERSATIONAL STYLES

It is important to remember that there are many and considerable individual differences between conversational styles. Tannen, D. [1991] suggested that males and females use conversation differently: males communicate information and facts with their conversations focusing on preserving status and avoiding failures, whereas females are more interested in swapping details of their lives and relationships, their conversations focussing on negotiating closeness and reaching a consensus. It is interesting to speculate that engineers, being mainly male, may be a reason why early AAC devices were focussed more on the male conversational needs of information exchange rather than social interaction.

Different cultures also have different styles—in some cultures it is impolite to say "no" and there are significant differences in turn taking behavior—some cultures expecting a pause before the next turn (e.g., Japanese and Finns) whilst others (e.g., New Yorkers) will tend to speak over the previous speaker. Birdwhistle, R. [1974] found substantial difference in conversational behaviour between a Pennsylvanian Dutch family (who talked on average 2.5 minutes per day) and a Philadelphian Jewish family (6–12 hours per day). This researcher claimed that equal amounts of new information were customarily transmitted in both families! There are also major differences in individuals' conversational styles within a single culture. Millen (in [McGaffey et al., 1991]) who was an AAC user, called for clinicians to be sensitive to (her) age and specific interests. Individuals will also use different styles depending on their conversational partner (their family and friends, their teacher, a policeman, and a very high status person, for example). An AAC user should be able to take different roles—e.g., requests, orders, friendly comments, annoyed, etc.

It was becoming possible to implement these ideas as computers were able to store increasing large amounts of conversational material. The challenge was how to ensure that users would be able to remember what was stored and how to access it [Light et al., 1990]. There was a need to find ways of grouping together and coding a database of phrases and sentences in a way that corresponded to the patterns of natural conversation. This led us to a major research program in which a variety of techniques were examined and developed. Many of these were either made commercially available via licensing agreements and the more generic ideas appeared in a range of commercially available AAC devices.

5.6   THE USE OF SPEECH ACT THEORY

The first useful concept we found was the "speech act" (an utterance which itself constitutes an act, rather than a transfer of information), and we discovered a number of useful guidelines for building models of conversation [Goffman, E., 1981, Gumperz, J., 1982, Schegloff and Sacks, 1973]. We found that conversations followed a ritual: the next utterance in a conversation is constrained by such factors as relevance, implicature and the "game" which is being played out by the partners [Levinson, S., 1983, Stubbs, M., 1993]. For example, the speech acts within an encounter between conversational partners A and B are normally:

A	Recognition—verbal salute—personal enquiry.
B	Response to personal enquiry and/or personal enquiry about A.
A	Response to personal enquiry.
A or B	Small talk.
A or B	Optional information exchange.

These are concluded by "departure routines" which also follow a predictable pattern.

Actual information exchange is only optional. Many conversational encounters, such as meeting a friend in the corridor, do not contain any exchange of real information. Their purpose is to acknowledge the other person's existence and to give an indication of continued positive regard. AAC users rarely participate in this activity, because it is perceived not important enough for the time it would take using a traditional AAC device.

We produced the simple and more complex models for a conversation that are shown in Figure 5.1. Although these are too rudimentary to be of much use in conversational analysis, we found them to be very useful for our particular requirements. This is similar to our experience in TV subtitling (see Chapter 3) and is another example of a sub-set of data from another discipline being useful in the AAC field. We chose the opening and closing stages of conversation as particularly suitable starting points for our work, but also incorporated 'fillers,' feedback remarks, and backchannel sounds. These speech acts contain no intrinsic information, but are important elements of speech and occur very frequently in every sort of dialogue [Edmundson, W., 1981, Goffman, E., 1981]. They allow the speaker to 'stall for time' whilst working out what to say next, hold the floor when they have not yet finished, and keep up a conversational rhythm [Beattie, G., 1979]. Examples of fillers are "Uh-huh", "I see", "Well, well". Some social groups use redundant expletives for this purpose. Feedback remarks include 'Agreeing' ("Yes I am with you on that"), 'Evaluating as good' ("that's really good" / "oh good"), or a request for more information ("Can you tell me more about that"). Backchannel sounds are used to indicate that a person is listening—these can be words, but are often just appropriate noises [Alm et al., 1988]. Yngve, V. [1970] has pointed out that conversations tend to be awkward without feedback of this sort. Interestingly, Maltz and Borker [1982] found that women gave more listening responses than men, so again this lack in AAC could be due to some extent to a male dominated development culture. These gambits could be particularly important for an AAC user of either gender, and could significantly improve their participation and control of an interaction.

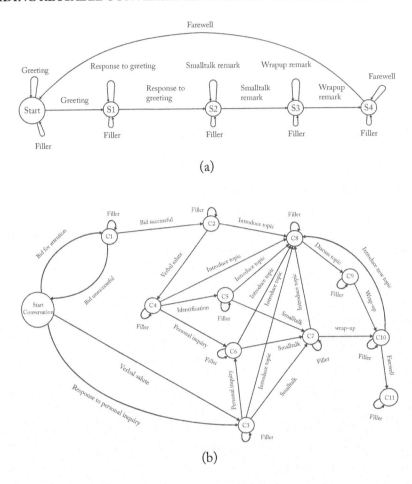

Figure 5.1: (a) A simplified model of conversation; (b) a more complex model of conversation.

5.7 CHAT—CONVERSATION HELPED BY AUTOMATIC TALK

On the basis of these ideas a prototype AAC device was developed called CHAT (Conversation Helped by Automatic Talk). This included:

- a simple dialogue model;
- a set of pre-composed stored phrases representing a wide range of conversational acts;
- a simple and highly efficient user interface; and
- an editor to create, alter and delete phrases.

The system (which would become a sub-system in a full AAC device) was designed to allow an AAC user to choose an appropriate phrase by a single keystroke that would cause the system to output an appropriate phrase or sentence. Such a system could significantly increase the speed at which an AAC user could say something. The phrases would need to match the personality of the user but, as they may be used often within a conversation, a particular phrase could become boring for the listener. We thus provided the ability for a user to store a number of phrases representing each speech act, the system randomly choosing one of these phrases. A further characteristic of conversations is the mood of the user. This tends to remain relatively constant during a particularly conversation, and we thus enabled the user to select a range of moods (polite, informal, humorous, and angry). Within each category of speech act a range of phrases were stored with appropriate phrases for each mood. CHAT also allowed a conversational partner's name to be inserted within these phrases, and the mood setting could also be linked to that name. To conduct a conversation the AAC user would initiate a greeting with a single keystroke. The system then automatically moves on so that the next utterance would be a response to a greeting, and so on through the conversation. Figure 5.2 shows a mock up of the visual displays developed for CHAT. Figure 5.2(a) shows the user being able to chose from a range of responses. The user has chosen key 5—"disagree", and has been offered the phrase shown in the bottom of the screen. Figure 5.2(b) shows how a simple conversational model has been incorporated into the control screen. In the situation shown, key 1 will cause the system to output a polite response to a greeting, including the name Steve, and key 2 will cause a filler remark to be outputted. Pressing key 1 also causes the highlighter to move to "small talk". This will mean that a single keystroke can now cause the system to output a polite piece of small talk appropriate to Steve.

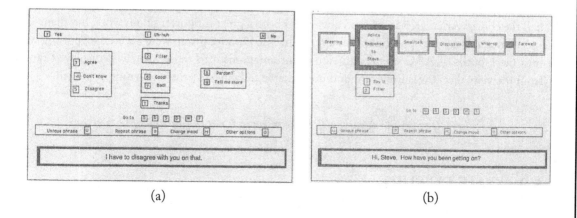

(a)   (b)

Figure 5.2: Screens from CHAT. (a) Giving Conversational progression; (b) the screen for "backchannel" remarks.

A number of non-speaking people used the prototype in early evaluation of this technique with very positive reactions [Alm et al., 1989]. The conversations appeared natural, and inappropriate remarks did not bring the dialogue to a halt. We had thus produced an efficient system for outputting opening and closing remarks. What was then required was to apply a similar technique to the central part of conversation—topic discussion. Arnott et al. [1988] developed a simple version of this that allowed the user to select a pre-stored sentence or 'story' on a particular topic, via a topic name search, but much further research was needed to produce an effective mechanism to allow AAC users to discuss chosen topics.

5.8   PRAGMATICS VERSUS SEMANTICS

We were essentially examining the importance of pragmatics in AAC use versus semantics (Penny Parnes—private communication 1988). Pragmatics focuses on the social context of language use and the conveying of attitudes beliefs and emotions of the speaker to the listener [Morton, K., 1992]. Our view was that the pragmatic goals of achieving participation in an interaction, and expressing the individual's personality throughout that participation, were at least as important as information content for much natural conversation, and this should be reflected in the design of AAC devices. CHAT and other similar systems were trading fluency for accuracy in an acceptable way—that is the users have less control over the exact words they say but can speak them more quickly and timeously. In addition CHAT encouraged phatic communication—phrases and sentences which conveyed little information, but were used to indicate closeness between people. Text communication and Twitter provide other examples of how modern technology is increasingly being used to provide phatic communication electronically for geographically distributed people.

These speech acts were not imposed on the user, they could revert to spelling a particular phrase, but this reduces the speaking rate significantly. The users were thus given the opportunity to balance control of the exact words against creating a sentence more quickly. Critically the user always had the choice. Arlene Kraat was and always has been a great supporter of our work, but many other speech therapists were opposed to this approach. The Dundee group were accused of 'putting words into the mouths of AAC users', and significant antagonism to these ideas was displayed ranging from the view that it was irrelevant and unworkable to being an ethically wrong approach.

If everyone agrees with it the research is not radical enough.

5.9   THE FUTURE

In the fullness of time, negative attitudes toward the use of pre-stored phrases changed, and CHAT like systems were subsequently introduced into many commercially available AAC devices. Researchers continued to investigate other ways to improve navigation through pre-stored data in AAC devices.

The use of pre-stored phrases may also have a place within text messaging and tweeting. Messages sent on these systems are often simply a way of keeping in touch with friends rather than sending important information. The range of methods that have been developed to increase the speed of communication for A.A.C. users could well be beneficial for users of these mainstream communication systems.

CHAPTER 6

Story Telling and Emotion in Synthetic Speech

Word-by-word AAC systems do not encourage users to make long statements. The concept of pre-stored phrases was further developed in the design of systems that facilitated extended conversations that were appropriate for a range of situations including story telling and giving lectures. Most AAC systems used synthetic speech, and we investigated how the prosody of synthetic speech could be controlled to allow emotional tone to be signaled within a spoken message. By the beginning of the 21st Century, many commercially available AAC devices included word and phrase storage and offered some control of the prosody of the speech output.

6.1   WHAT DO WE DO AFTER WE HAVE SAID "HELLO"

Story telling permeates non-aided conversations and provides a way of forming and making sense of people's experiences and relating them to each other [Quasthoff and Nikolaus, 1982]. The fact that we tend to reshape and embellish these stories indicates that pragmatics can over-ride other considerations. Non-aided speakers tend to tell the same stories to a number of different people, with only minor modifications (Fillmore's 'rehearsed language' quoted in Gumperz, J. [1982]). Such narratives fall into a number of categories, some will be short lived, (holiday experiences) or constantly up-dated (family news), but others are more stable and used over much longer periods (biographic details, hobbies). Few stories are only told once. This provides a more wide-ranging concept of the inclusion of re-usable conversation within an AAC device. CHAT type systems facilitated communication, but they did not include any way in which an AAC user could easily tell stories. We thus examined a number of ways in which we could include re-usable stories within an AAC device.

"We are all stories—just stories."

Our approach was for the user to select a topic, conversational partner, and speech act, rather than by typing the works they wish to say. The system then searches a database for an appropriate utterance based on these characteristics, and offers the results for the user to decide whether or not it is appropriate. These systems were developed for users who retained language comprehension.

6.2 TOPIC—TEXT OUTPUT IN CONVERSATION

TOPIC (Text OutPut In Conversation) [Arnott et al., 1988] consisted of a database, database management software, and an intelligent user interface. This was an unusual application for data base software, and we used a novel (at that time) text database, which allowed searching via speech acts, topic description, specific words, and/or frequency of use. TOPIC offered a window of predictions of appropriate next pieces of text

Waller et al. [1991b] applied the techniques of artificial intelligence to sentence and conversational narrative prediction using social knowledge—jokes, family news, holiday stories. They developed Talks:Back (Talking and Learning Knowledge System for Better Aphasic Communication) [Broumley et al., 1990]. TalksBack included a semantic network to reduce the cognitive load traditional database search tools imposed on the user. This represented the users' personal and world knowledge—e.g., their network of friends and their interests. The user enters information about themselves, their conversational partners, topics, social situations, and speech acts into the database. In the Conversational Mode the user sets the scene by selecting a subject, partner, speech act, and, if relevant, situation. The system selects possible utterances from its database, and the user can either choose to speak the utterance, or request that the search be continued. Figure 6.1 shows a screen shot from TalksBack, which was successfully piloted with four dysphasic clients. This research emphasized the need for a supportive and motivated primary carer for such systems to be successful [Waller et al., 1998].

Although developed as separate projects, the modules—PAL, CHAT, Topic and TalksBack were designed to be incorporated into a multi-level integrated aid. This research was commercialized by Don Johnstone as the softwarepackage Talk:About, screen shots from which are shown in Figure 6.2.

6.3 PROSE—AND OTHER STORYTELLING SYSTEMS

Prose [Waller et al., 1991b] was an extension of the TalksBack concept designed to reduce the cognitive load of storytelling. It predicted possible texts and allowed the user to navigate through the text of the story, as well as offering ways of pausing, interrupting and modifying text in mid-conversation. Prose was used by a woman with expressive aphasia and limited cognitive abilities who was reported as an excited and enthusiastic user.

McKinlay and Newell [1992] had introduced conversational analysis to the AAC research community in a research symposium in Philadelphia, explaining how it would enable designers to better understand the structure of conversations. In particular, they examined how people managed conversations [Garfinkel, H., 1967] and the turn taking mechanisms that are used in non-aided conversations.

Taking non-speakers' social difficulties as a starting point, McKinlay, A. [1991] examined how he could instantiate these in an AAC system. His system was called MOSCO (Modular Social Communicator). After the user had uttered a phrase, the system searched its database of phrases

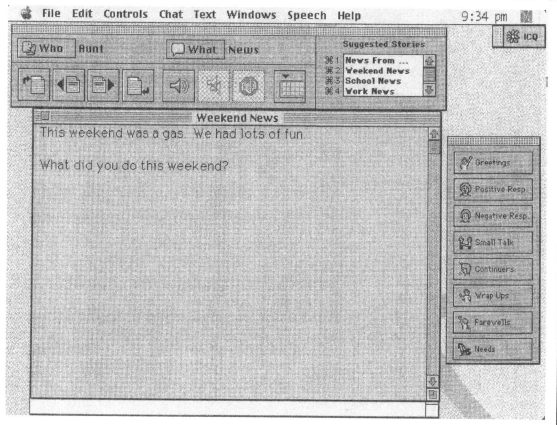

Figure 6.1: TalksBack—an early prototype.

to find a number of suitable successor phrases using semantic analysis. The user chose the one they found most appropriate for the situation. An evaluation of this device underlined the importance of working at a sentence or phase rather than word level in order to increase perceived communication competence.

Todman and Alm [1994] produced and experimented with a paper version of a topic discussion generator (Todman et al. [1995]) and followed this by a Hypercard simulation where a non-disabled researcher communicated with eight conversational partners with very promising results. Following this work they developed TALK (Talk Aid using pre-Loaded Knowledge). Content was organized on the basis of the person (my versus your), time (past, present, future), and orientation (where, what, how, when, who, why). This also included fillers—plus the important "repair fillers" for when the conversational partner has misunderstood what is being said. Figure 6.3 shows an example of a screen shot of TALK. A version of CHAT can be seen across the top of the screen. The panel

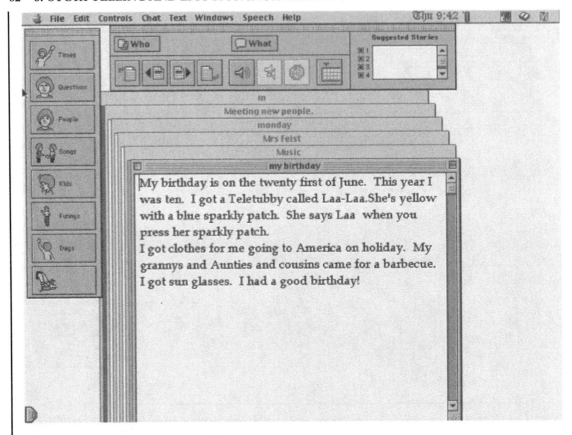

Figure 6.2: Talk:About.

on the left is used to access the database for sentences about the user ("me") or the conversational partner ("you"), and the type of sentence ("where", "what", etc.). Sentences drawn from the database are shown in the center panel. The right-hand panel allows the user to access feedback remarks.

Todman et al. [1995] describes an evaluation of such a system with a non-disabled user. These showed that social competence rating of transcripts of conversations were higher with the use of TALK than unaided conversation. The joint work between the School of Computing and Todman was licensed to Mayer Johnson Co, PO Box 1579 Solana Beach California, CA 9205 U.S., who marked the system as "TALK BOARDS".

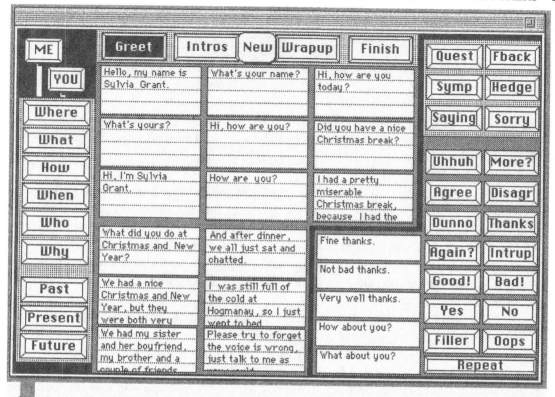

Figure 6.3: Hypercard simulation of Talk.

6.4   FLOORGRABBER: AN AID FOR NON-SPEAKING LECTURERS

Floorgrabber was developed in a cooperative effort between the researchers and Alan McGregor a non-speaking person who was a part-time member of our research team [McGregor et al., 1991]. McGregor has been an invaluable member of the Dundee research team since he joined it as a school pupil. He comes into the University one day per week and has played an important role as a "user" and by stimulating ideas within the team for over 20 years. He has acted as a valued colleague and ambassador at many international conferences. At the 1992 AAC conference in Philadelphia, McGregor and Alm [1992] used Floorgrabber to describe his role in the paper "Thoughts of a nonspeaking member of an AAC research team".

Disabled users as valued members of research teams

Alm et al. [1992] designed Floorgrabber as a conversational and lecture aid for Alan McGregor. It used hypertext and incorporated stories which the user could access in the future (e.g., a number of paragraphs were prepared based on a journey abroad to an international swimming competition). The system enabled the user to choose which paragraphs to use in any conversation, and in which order to use them. CHAT—like quick fire quasi-randomly selected remarks were also available.

An extension to this system was developed using ideas from "fuzzy" information retrieval [Negoita, C., 1976]. This allowed the system to find the closest matching set of candidate utterances for the next speech act. In contrast to conventional systems, the fuzzy set system always produced a full set of candidate utterances, and reduced the number of mouse clicks required for a conversation [Alm et al., 1993]. The 'Fuzzy talk' system also modeled, in a simple but elegant way, the "step-wise" progression from topic to topic which marks natural conversation [Jefferson, G., 1984].

Floorgrabber was formally evaluated in a single case study experiment. Substantial increase in words spoken and an improved conversational control were found compared with McGregor's 400 word normal Bliss board system. McGregor used this system in many national and international lecture tours. He gradually expanded the database of Floorgrabber by adding replies to the questions he was asked at each of these lectures. In this way, McGregor built up a system so that could be used successfully during most of the question and answer sessions of his lectures.

6.5 OTHER CONVERSATIONAL AIDS

Script Talker was developed as part of a European TIDE project [Dye et al., 1998]. The message database was displayed on the screen as being within a virtual town, the users being able to move from building to building. When entering a building, the user is presented a scene associated with it (such as a restaurant interior) and messages associated with objects in the scene are presented and may be activated by pressing on, for example, a glass, a menu, or the bill. Figure 6.4(a) show a CHAT and feedback remarks screen, using icons to indicate the type of phrase that will be uttered. Figures 6.4(b) and 6.4(c) show the screens that would be used for conversational interaction in various situations. Other similar systems that were developed at that time included CAMELEON, part of the TIDE VASS project and Profet developed by KTH and reported in [Newell et al., 1998].

Research using database approaches included the development of WordKeys [Hickey and Page, 1993] based on research into intelligent text retrieval. The system, shown in Figure 6.5, automatically indexed new messages and stored them, together with the index, in a database. They were retrieved by typing in one or more index words, or words with a similar meaning to the index words. SchemaTalk, described in Vanderheiden, P. [1995], provided

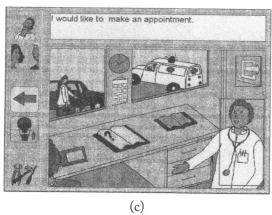

Figure 6.4: Script talker. (a) A CHAT screen; (b) in a shop; (c) at the Doctor.

easy access to message by organizing them in a database using a hierarchical schema approach. CONTACT was an utterance based prototype developed jointly by the Universities of Buffalo, Dundee and Abertay Dundee, and Enkidu research which included quick turn taking structured interactions, informal chat and text entry with word prediction. Higginbotham et al. [2007] compared the performance of a number of systems, including "TALK", "FRAMETALKER", and "CONTACT".

A predictive communication system for prompting people with Broca's aphasia in their communications, was developed by Waller et al. [1998]. The systems database was populated with personal sentences and stories by a caregiver. The user retrieved these conversational items using a simple interface which offered probable items based on previous use of the system. Five adults with

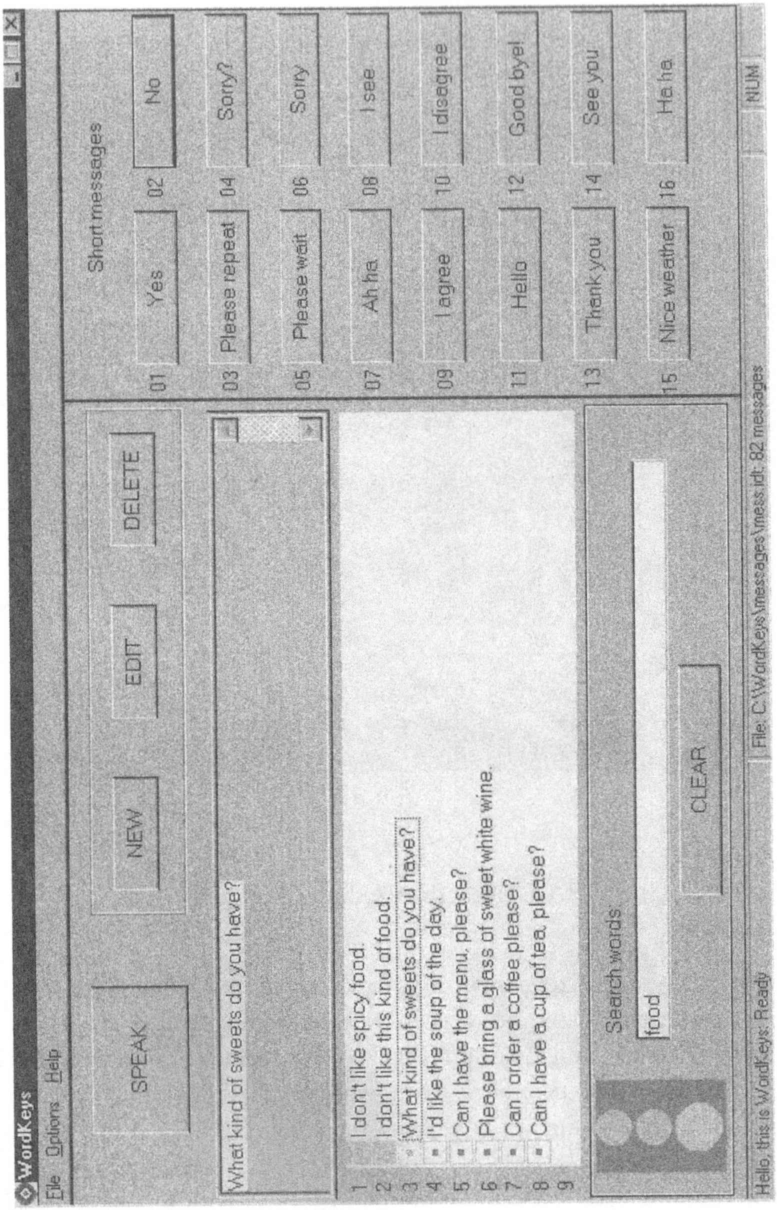

Figure 6.5: Word keys.

non-fluent aphasia, who were able to recognize but not produce familiar written sentences, were able to initiate and retain control of the conversation to a greater extent when familiar sentences and narratives were predicted. This project indicated that a communication system based on prompting could help people with cognitive and communication difficulties. A similar approach, described in Chapter 8, was adopted many years later by Alm for people with dementia.

6.6 JOKES

Waller, A. [2006] examined ways of using natural language generation to provide opportunities to extend both vocabulary and the type of conversation in which children with language disorders can engage. The STANDUP project [Waller et al., 2009] provided technology that generated puns. In an evaluation of this system, nine children with cerebral palsy were all able to generate puns; this also increased the children's ability to categorize words suggesting that the ability to tell puns can have an impact on language abilities. A further indication of the educational value of AAC devices is shown by the "How was school today?" project [Black et al., 2010, Reiter et al., 2009]. These authors produced a working prototype, using location data from RFID sensors in the school, together with timetable data, and recorded messages from teachers, to generate stories for future conversations. A narration interface was designed to support interactive storytelling, allowing the child to narrate events in any order and to add additional evaluations. Results showed that the system supported storytelling skills and memory sequencing, and enabled users to engage in enhanced interactive conversation.

6.7 SYMBOLIC AND PICTURE-BASED COMMUNICATION SYSTEMS

The developments described above are text-based systems, but many non-speaking people have poor or non-existent written language. Much research had been focussed on either symbolic or picture based communication systems as an alternative to text; see Beukelman and Mirenda [1998]. These include Picture Based Communication Symbols [Johnson, R., 1981], Picture Ideograms [Maharaj, S., 1980] and Rebus [Clark, C., 1984]. Symbols were used both on communication boards and electronic systems such as Speaking Dynamically (Mayer-Johnson Co., U.S.) and Talking Screen (Words+ Co., U.S.). The users of such systems, however, tended to produce telegraphic messages [Bruno, J., 1989], and successive single word sentences with no syntactic organization [Kraat, A., 1991] .

One of the first AAC symbol systems used Bliss Symbols rather than pictures. Bliss [1965] developed these as an international language "which would lead to world peace" [McDonald, E., 1980]. Blissymbolics is a semantic-based natural written (graphic) language, with a structure similar to Chinese. McNaughton and Kates [1974] discovered Blisssymbolics and had the vision to realize that it had potential for use by people with complex communication needs. Bliss sentences are created using a sequence of one or more Bliss characters, and thus predictive algorithms can be applied to

Blissymbolics to assist users in the retrieval of words. A number of electronic Bliss systems were developed including the BlissWord project [Andreasen et al., 1998], which investigated ways in which Bliss users with physical and cognitive limitations, who may not be literate, can explore new vocabulary.

6.8 COMMUNICATION AIDS FOR INTENSIVE CARE

All AAC devices should have as intuitive an interface as possible, but an extreme case of the need for an intuitive interface is within Hospital Intensive Care (ICU) units. Patients may be on ventilators and thus unable to speak, and they will be very ill. Many patients are frightened of the ICU environment and frustrated that they cannot 'speak'. Their lack of effective communication means it is difficult to tell anyone about their anxiety during this period of intubation. Alphabet boards, pen and paper, or mouthing words are time consuming and frustrating for both patients and nurses, and many patients describe feelings of disempowerment and social isolation whilst intubated [Etchels et al., 2003]. Any AAC system in this environment, however, would have to be useable with minimal or zero training.

Ian Ricketts led a collaborative project involving the Department of Applied Computing, the School of Nursing and Midwifery at Dundee University, the Department of Speech and Language Therapy and the Intensive Care Unit at Ninewells Hospital, Dundee. They examined the communication problems of these patients, and developed an AAC device called ICU-Talk [Etchels et al., 2003]. This was based on the TalksBac system, phrases or questions could be selected from a pre-stored database, but had a very simple and intuitive interface that required minimal training. Figure 6.6 shows an ICU-Talk installation, and a sample control screen.

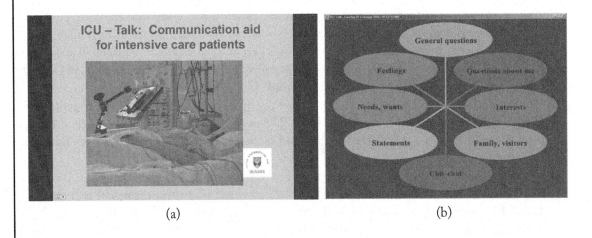

(a)   (b)

Figure 6.6: AAC in Intensive Care. (a) ICU-Talk; (b) output screen of ICU-Talk.

The vocabulary for the ICU-Talk database was collected by asking nursing staff from an intensive care unit to give examples of communication attempts by patients. This was followed by an observational phrase, where patients were observed for a 36 hour period. The examples given by the nurses were compared to actual communication attempts by the patients and approximately 50% of the attempts were found to be specific to a particular patient. A computer-based interview was used with close friends or relatives to provide personalized phrases and questions for the data base. This was a very useful part of the process as some of the conversational topics suggested came as a surprise to the nursing staff. This underlined the importance of an aid being personalisable—even a relatively simple aid.

Nineteen patients used ICU-Talk for up to six days with a mean length of use of two days. The single case studies with ICU-Talk suggested that this device can be used by intubated patients to augment their attempts to communicate: none of the patients indicated fear or uncertainty of using it.

6.9   EMOTION IN SYNTHETIC SPEECH

Very early AAC devices used printed or a visual output but, as soon as synthetic speech achieved a reasonable quality, it began to be used in AAC devices. Although most speech synthesis research had not been directed at use with disabled people, the major use of early speech synthesizers was for AAC devices and devices designed for blind people. One of the disadvantages of early AAC systems was the restricted nature of speech synthesis—early speech synthesizers offering only a male American voice with a neutral emotional tone. Alm and Newell [1996] discussed their concerns that a limited range of intonation can imply a corresponding limit to the emotional range and thus intelligence of a user, and an eminent AAC researcher once commented that "sex is more important than intelligibility". Ideally, the voice should also match the socio-economic and ethnic characteristics of the user, as well as their mood.

> *"It ain't what you say, it's the way that you say it"*.

Iain Murray pioneered the introduction of emotion into synthetic speech. Emotion is conveyed in natural speech primarily by sound quality and intonation patterns, and Murray and Arnott [1993] provided a review of the literature on human vocal emotion. It is clear that the ways human beings introduce emotion into their speech are very complex [Frick, R., 1985]. Again, however, a gross simplification of what was known in this academic field proved useful for our application area. HAMLET [Murray et al., 1988] was designed as a module that could be added into a conventional speech synthesizer (the DECtalk speech synthesizer kindly donated by Dave Lawrence of Digital Equipment Corp Ayr). It modified the output of the synthesizer to simulate six different emotions (happiness, sadness, anger, fear, grief, disgust). These emotions were chosen as they were the most thoroughly studied in the literature, and thus information on their acoustic features was available.

HAMLET modified voice parameters, and prosodic rules to alter the pitch and timing of the output speech. In a listening experiment, all emotions were perceived best when used with semantically appropriate text, although anger and sadness were also often recognized out of context. The use of a joystick to control emotion in the two/three dimensions described by Scholsberg, H. [1954] was also considered, but had the disadvantage that it could involve too complex an interface for most non-speaking people. Cahn, J. [1988] did some contemporary research at MIT for an "affect generating system" [Cahn, J., 1989] rather than as part of an aid for non-speaking people.

The meaning of even a simple utterance is dependent on the tone of voice as well as the word spoken. "Yes" can be an affirmative response to a question, but it can also be used to indicate sarcasm, disbelief, tentativeness, or even "no". Crystal showed nine different intonation patterns for responding "yes" to a question, but this only touched the surface of the richness of expression that most speakers employ [Pullin, G., 2009, p. 159]. Pullin, G. [2009, p. 165] quoted Paul Johnson as saying "no man is truly English if he cannot say 'really' seventeen different ways."

Graham Pullin, an engineer and creative designer, joined the Dundee group. He had a deep interest in phonetics and AAC devices. He was particularly interested in the balance between semantics and pragmatics within conversation. He built a device that could produce only two words—"yes" and "no", but various ways of controlling the intonation pattern of these words were built into the system. Together with Waller and other colleagues, he developed the Phonic Stick in which a joystick was used to control the output of a speech synthesizer, with the intention of enabling non-speaking children to experiment with making speech sounds (as speaking children do) rather than speaking preformed words [Black et al., 2008].

Pullin approached the design of AAC devices from the perspective of a rehabilitation engineer turned creative designer. He suggests in his book "Design Meets Disability" [Pullin, G., 2009] that rehabilitation engineering and assistive technology should make more use of the design techniques that creative designers have developed. This approach is proving very fruitful in his research into AAC and has highlighted a number of fascinating and important research issues in this field [Pullin and Cook, 2010].

6.10   THE CURRENT SITUATION

Higginbotham et al. [2007] provided a review of the AAC field in 2007. He commented that, "in recent years, a social interaction paradigm has begun to affect research and development activities. This sees human communication as a joint action (negotiation), and includes perceptions of communication competence, understanding social interaction as cooperation towards joint goals, and the inclusion of facial and other gestures and body language". This resonates with the comment in Alm and Parnes [1995] that "a good AAC system will allow for intonation, facial expression and gesture and encourage the graceful combination of the AAC communication with the speaker's other modes".

Higginbotham concluded that the emerging interface options include:

- adaptive scanning (rate adjusted based on user performance);

- "so-called" "brain interfaces"—which are still at experimental stage;

- automatic recognition of dysarthric speech;

- gesture analysis & multimodal input;

- visual screen displays—he reports on evaluation of the effectiveness of these and what needs to be improved.

He also comments that speech synthesis lacks inflection control needed to construct pragmatically appropriate utterances.

Judge and Towend [2010] conducted an extensive survey in the UK involving interviews with 18 AAC users, with paper and on-line questionnaires being completely by (or on behalf of) 43 further users and 68 professionals (mainly speech and language therapists, but also other professionals such as special education teachers). The results show that current devices were not considered reliable or durable. Many of them were difficult to use resulting in a high cognitive load particularly when large vocabularies were involved. Complaints were also made that, for example, in some cases a helper was needed to switch the device on and off.

The highest ranked need for improvements, were:

- speed of communication;

- ease of use;

- usable everywhere;

- being able to access vocabulary quickly and easily; and

- being able to personalize the device.

Users and their carers reported feelings of frustration, anger and panic when they were unable to rely on their device working well, or were left for long periods of time without a working device. These suggestions are an indication of the rate of progress in this field over the past 20 years, and how much more needs to be achieved.

6.11 THE FUTURE

The results of the research described in this chapter now appear in many commercial devices, either licensed or from generally available information published in the academic press. Of the modern commercially available AAC devices, *TANGO* (http://www.Blink-twice.com) is an example of an aesthetically pleasing device which contains many commercial innovations a number of which are direct descendents from the research described above [Pullin, G., 2009, p. 175].

Kraat, A. [1990] spoke of the need of AAC devices to "achieve a real and meaningful change in the communication/social interaction and lives of persons requiring AAC". I followed this up in my Phonic Ear lecture [Newell, A., 1992b] where I advocated a "paradigm shift" in thinking about AAC—a shift which Pullin, G. [2009, p. 161] claimed has yet to take place.

> *AAC technology needs to create real and meaningful change.*

An ideal to aim for may be truly invisible systems that learn to facilitate the preferred conversational patterns of users. In the meantime, in terms of Kraat's "real and meaningful change", the reports of AAC users demanding high levels of reliability in their AAC devices and becoming very angry when they go wrong are perhaps an indication of a device which goes someway towards Kraat's "real and meaningful change" criterion.

CHAPTER 7

Lessons Learned from Designing AAC Devices

The development of AAC systems described in previous chapters led to many insights into the most appropriate way of conducting research into this and other similar fields. Lessons learned included the composition of the design team, the relationship between developers and potential users of assistive technology, and the ethical issues, and potential conflicts that can occur in interdisciplinary research with a clinical context.

7.1   AN EFFECTIVE AND APPROPRIATE TEAM

Assistive technology should achieve a "real and meaningful change" in the lives of disabled people. In order to do this, designers and researchers need to have an appropriate team that includes:

- engineers with empathy and insight and imagination;

- clinicians and other relevant disciplines;

- a good relationship with potential users;

- links with clinical practice; and

- an understanding of the lives of the people for whom one is designing.

The team should have:

- an appropriate research methodology; plus

- a knowledge of the literature; and an awareness of current commercially available devices, and crucially,

- some good ideas.

The composition of the research team is of major importance. Clearly, an engineering development needs to have people with appropriate engineering expertise (software and/or hardware depending on the project). Unlike many mainstream engineering projects, however, these engineers may never have had any real contact with people who are potential users of their systems. It is vital that the engineers can relate to the potential user group, and are able to develop an insight into their needs and wants and the challenges of providing technology to assist them. The engineers need to imagine themselves in the situation of their potential users, in a way that does not patronize these people. It is very valuable for the engineer to have contact with clients in realistic environments. Sitting in on school lessons for children with special learning needs, for example, can be very valuable. This, however, can raise ethical issues and requires a careful structuring of research and development projects to ensure that the appropriate training and support is given to the engineers. Ideally the design team should contain at least a therapist, or other appropriate clinician, and one or more potential users. It is important that the therapist/clinician realizes that they have a research rather than a therapeutic role—their main role is not to "help" the individual potential user, but to use their skills to design a future aid for a range of users in the future.

7.2   THE ROLE OF USERS IN THE DESIGN TEAM

It needs to be made absolutely clear to potential users in the team that the research is not primarily aimed at benefitting them personally—they are there to help with the project, and any help the project is to them is secondary. These disabled consultants provide ideas and comments and act essentially as test pilots for prototype systems. Where appropriate there should be user panels who take part in focus group activities and who can assess and evaluate systems that are produced as part of the research, either via focus groups or as part of case studies using such systems.

Users on the research team should have appropriate rewards. These need not necessarily be monetary, as such rewards, other than travel and similar expenses, may interfere with state benefits and/or pensions that the users may be receiving. Thought should be given as to how to make such rewards appropriate and appreciated by the users. In one case the Dundee team wished to employ one of their user group, but did not as that may have caused the user's benefits to be cancelled, and it could have been difficult for the user to have those benefits re-instated: (grounds for refusal being that they had "shown themselves capable of work"—even though the work in the research project was very specialized and unlikely to be available in any other situation). Users and clinicians can also be focussed on incremental improvements, and, as has been reported in Chapter 5, some can be positively hostile to novel ideas. They can also view research in the field of assistive technology, especially in its early stages, as being seen as an unnecessary diversion from the task of improving the current situation.

We were fortunate in employing engineers with cerebral palsy working full time on research projects: Annalu Waller (a rehabilitation engineer) and Greg Filz (a severely cerebral palsied software engineer) (see Peddie et al. [1992]). A colleague commented about Filz that: "there is nothing that focuses the mind on reality as having lunch with a non-speaking researcher with cerebral palsy whom you know to be a far better computer programmer than you."

Disabled specialists can play an invaluable role in research.

It is vital that all members of an interdisciplinary team fully understand and openly appreciate the contributions and constraints of other members of the team. The clinician needs to understand that some tasks are easy for a computer and some impossible, and the engineer must focus on the real needs of the situation rather than "interesting technical challenges". The therapist does not need an engineering training or vice versa, but there must be mutual respect. They should know enough about each other's disciplines to communicate effectively and understand each other's jargon. They need to be prepared to acknowledge their lack of expertise and realize the value of bringing together knowledge from disparate sources rather than a very deep knowledge from a single discipline.

7.3   RESEARCH IN A CLINICAL ENVIRONMENT

A further challenge with assistive technology research is the relationship between research and clinical commitments. It can be particularly difficult in a clinical situation to make an appropriate distinction between research and service commitments; researchers in clinical environments need to allocate an appropriate proportion of high quality time to research and not allow routine clinical commitments to encroach upon it. The balance between maintaining vital clinical contacts and day-to-day problems is a very delicate one for both clinicians and engineers [Newell, A., 1987a]. A separate research laboratory can be very helpful, together with specific time periods allocated to research that can be interrupted only by emergency situations in the clinic. The Dundee group are in an academic research environment. They do not have any commitment to service provision but work closely with local therapists and schools, individual users, and groups of users. Our challenge was to ensure that the clinical contacts we had were adequate for the research. Clinical colleagues need to be chosen with great care. The team needs to take full advantage of their expertise, but should not necessarily be dominated by their views. Clinicians may have conflicts of interest between satisfying their needs and those of the clients. For example, in our work with intensive care patients, nurses suggested that the patient would wish to concentrate on their medical condition, whereas we found that patients also wanted to discuss social and family matters. In a more extreme case, a 'silent' patient can have advantages for the clinical team, and there have been rumors of non-accidental damage of AAC devices—or devices being switched off by relatives, or left beyond the reach of the patient. The on/off switch of a communication aid should be accessible to the user!

It is important not to focus entirely on what the disabled user cannot do. There is a need to examine what the disabled users can do and how to build on these abilities. Gajos et al. [2008] showed how abilities can be accommodated within an ability-based design for motor impaired users, and the "Ordinary and Extra-ordinary" concept (see Chapter 9) encourages designers to focus on the abilities as well as the disabilities of potential user groups.

7.4   COMPOSITION OF A RESEARCH TEAM

The research team should include the disciplines which are relevant to the research, and the multi-disciplinary Dundee group have been fortunate over the years to employ on research projects—speech and occupational therapists and a special education teacher, linguists, psychologists, conversational analysts, media (film and theatre) professionals, and creative designers. We have also been able to employ people with a mixed background (examples include: my own academic background in electrical engineering and psychology, Peter Gregor, who has a background in psychology, computing and community education, Norman Alm, in English and social work, Anna Dickinson, history and computing, and Graham Pullin in rehabilitation engineering and creative design.

The challenge of leading such teams is being able to pull together the contributions from this wide range of disciplines, and not give too great a weight to the leader's own discipline. It can be a great advantage if such a team leader decides to become more of a generalist rather than a specialist, but this inevitably means that they will have a reducing knowledge of their own discipline. The greater reliance he or she will have to place on other members of the team to keep up-to-date with the advances in their own field will also be advantageous from a team perspective.

7.5   AN APPROPRIATE RESEARCH METHODOLOGY

A multi-disciplinary team is vital for research into assistive technology, and so is a User Centered Design Approach. Much has been written describing User Centered Design (UCD), Participatory Design, and other similar processes (see Chapter 11), but these recommendations rarely, if ever cover the special challenges of research and development of assistive technology. UCD often assumes that it will be fairly easy to recruit sufficient representative users. This is rarely the case in AAC, and it can be very difficult to find "representative" users for this type of research. The numbers of users in a particular location will not be high, and there may be practical and ethical difficulties in recruiting them for a research project, that could include the difficulties of obtaining informed consent for users who cannot communicate and/or who are legally "incompetent".

7.6   EVALUATION TECHNIQUES

The small number of users also raises the question of how to provide an acceptable evaluation. The medical field has an increasing focus on "evidence based medicine" which is very closely linked to the "gold standard" of double-blind randomized studies. This latter is well suited to pharmaceutical products, but much less applicable to assistive technology. How, for example, would a double-blind randomized evaluation trial be designed for the efficacy of a walking frame?

A further problem with evaluation is that older and disabled users' gratitude for the efforts the researcher is putting into the project, or simple (but misplaced) politeness, may mean that they communicate a much more positive view of the project than they really think. In our work, we often imply that the systems were designed by someone else in the hope that this will produce a more accurate picture of what the person actually thinks. Alm, N. [1994] describes how difficult it was to persuade AAC users to be critical about research prototypes. They had the greatest difficulty in thinking up critical comments let alone saying them from a public platform.

Assistive technology is seen by many as a medical intervention, whereas it is often more obviously a domestic product. The evidence base required for potential benefit of domestic products is much less demanding than for medical products: the automobile, the home computer and the mobile phone had little or no evidence base when they were first introduced. Perhaps we should reduce the barriers to the introduction of assistive technologies, because many are essentially domestic rather then medical products.

7.7   THE COST OF ASSISTIVE TECHNOLOGY

The cost of communication aids, and other assistive technology, is often considered to be too high, particularly those that are software based. Any product that can only be sold in small numbers, however, needs to be more expensive than a mass market product as the development costs have to be amortized over a relatively small number of sales. The cost can encourage a therapist to go to a technically competent friend and ask them to produce an aid, or to an academic colleague suggesting a student project. In these cases "cost" is usually estimated on the basis of materials used—rather than time and effort. Occasionally this has been successful, but what is not realized is that, if an aid is going to be useful, it must be robust and there is a need for long term maintenance, which friends or college students are unlikely to be able to provide. This approach underestimates the skills necessary to produce a good communication aid, implies an unrealistic cost structure and reduces the commercial viability of this market sector. Therapists would be unlikely to ask a friend or a local college to make an electric toaster for one of their clients, but some appear prepared to do this for communication aids! There are many disabled people who have been disappointed and have lost faith in technology because of the efforts of well meaning engineers who were badly advised by therapists or vice versa.

There is also a problem of the user of assistive technology devices seldom being the purchaser. In the UK the purchaser is often the National Health Service, or a charity, and in the U.S., an insurance provider, who will usually be advised by clinicians not users. This raises questions of conflicts of interest, and also creates the perception that the device is a medical rather than a domestic device. The design, or even the decision to attempt a design, may thus be fatally compromised by payment rules and regulations [Newell, A., 1987b].

7.8   PROFESSIONALISM IN RESEARCH

There can be a problem with the professionalism of some people who "dabble" in assistive technology research. Occasionally, in the UK at least, an academic will meet a clinician at a party, or some other event, and become fascinated by the challenge presented by a patient they describe. The academic then sets up a project to solve the particular challenge presented. At no stage do they question the clinician about the state of the art of assistive technology—often there is a commercially available device which will solve the problem, but the clinician is either not aware of its existence, or it is too expensive for their budget.

Similarly, researchers sometime do not to investigate the research literature, and there is a very great deal of wheel re-invention. My colleagues and I have had to point this out to authors on many occasions, and I once remarked that, although an idea described in an academic paper was a good one, there had been reports of such devices in the assistive technology literature over 40 years previously. It is difficult to understand why otherwise good researchers, who normally would not dream of ignoring the literature, do so when they are developing assistive technology—this wastes much time and effort and essentially patronizes users of assistive technology. There are too many challenges to address to waste our efforts on those that have already been solved.

7.9   ETHICAL CONSIDERATIONS

There is a whole range of ethical issues involved in developing AAC devices. A particular challenge is the different time scales—clinical service intervention has an immediate or few months time-scale, whereas the fruits of research may not be seen for a number of years. This is important when the sophistication and cost of technology is changing rapidly, and clinicians are not aware of potential changes. Early research into the use of computers for therapeutic/educational intervention was attacked on the economic basis of the costs of computers at that time. Later the same situation applied to mobile phones, and more recently iPads as the platform for therapeutic devices.

In the 1994 ISAAC research symposium, Alm, N. [1994] published a detailed discussion on the ethical issues involved particularly in AAC research, but many of the comments also apply more generally to assistive technology research. He made the point that, when dealing with non-speaking people, it is almost inevitable that the researcher will also be dealing with their clinician, helper or family. The researcher, however, needs to be aware of the power relationship between non-speaking user and their helper (and the potential for conflict of interest).

Todman et al. [1995] made a case for using non-disabled people for early evaluation of devices to avoid wasting the time of disabled users, especially when the use of any prototype system may put significant physical or cognitive strain on a disabled user. They suggested that "trials with people simulating non-speaking users may be regarded as (necessary) hurdles to precede the involvement of people with real communication problems". For complex systems the researchers need to ask whether is it reasonable to ask a person to spend many months learning to use a system that may never become available to them—particularly if the user has a degenerating condition. In addition, older and disabled people can have a poor self image and a lack of confidence in using equipment. Thus, although failure on behalf of the user can provide valuable research data, this should be avoided with most older and/or disabled users. This can be achieved by initial experiments with people simulating non-speaking users.

It should be emphasized that, at certain stages in the research, experiments with disabled people are essential. These have to be planned with much care to ensure that the subjects are fully briefed and their expectations are not unduly raised by these experiments. A particular problem can occur with student projects—the disabled person is given a great deal of attention that suddenly finishes when the student's period of study is over. With some disabled people this can occur every year (Murphy quoted in Alm, N. [1994]). With care and sensitivity, however, these challenges can be overcome and in general we have found that our users enjoy the experience of working with us.

In pioneering research, Prior et al. [2011] developed requirements gathering techniques appropriate for people with Complex Communication Needs who use AAC devices. These included both focus groups and group discussions about paper and other prototypes. These authors were careful to build up long-term relationships with their users, they employed multiple choice consent forms, and carefully planned the focus groups to be sensitive to the needs of the participants. The participants were fully engaged in the process and provided very useful information to the developers.

Alm, N. [1994] also discusses the problems that can be caused by press publicity of research projects. Clearly, the research institution is keen on such publicity, but such publicity may not make it clear that this is a research prototype and thus not (yet) commercially available. This can result in the researcher receiving many phone calls that result in disappointed users, or helpers. It is unrealistic to expect that all research will lead to a commercially available and supported product—and the research team should think carefully before giving up research to become a service organization (selling the product is only the beginning—it also needs maintenance, and possible training programs and a support desk).

The PAL project raised a number of ethical issues. We had been developing and evaluating PAL in collaboration with local schools, and had provided them with a small number of computers on which PAL could be run. These computers belonged to the research project and, when it ended, they had to be withdrawn from the school. Newell et al. [1992] report a teacher's comments: "This research has been so successful in every way that I would like to pupil to continue. I fear that he might just revert to being an isolate and "switching off" school work. Psychologically it seems cruel to withdraw a resource that offered positive help or, as he said himself, "it is too good to last". At this point in any project researchers cannot win. Either the research is a failure and withdrawing the computers is not a problem, or it is a success and withdrawing the computers causes distress in the participants (and their parents). Fortunately in the case referred to above, the education authority were persuaded to provide replacement equipment. We advised the schools and the parents of how they may obtain replacement computers, and PAL was made commercially available—which removed the danger of the schools using software that was not maintained (after the end of the project there were no funds in the university for maintaining software which has been developed). We also made sure that participants in our research realized that they were experimental subjects and we could not support them after the end of the project. One could suggest a measure of success of a project as the amount of trouble caused when the project ended!

Institutional inertia is another challenge faced when new assistive technology software is introduced. Although PAL was helping students greatly, for some time it was not allowed to be used in examinations, as it would "unfairly advantage the child"! (In our view this was the same as suggesting children with visual impairments could not wear their glasses in examinations). We lobbied against this view and eventually the decision was overturned.

It is valuable to talk to psychologists and therapists about ethical issues because appropriate professionals can have a detailed knowledge and experience of ethical issues that must be addressed but, as Alm, N. [1994] commented, "Perhaps the most important sensitivity is to realize that there are potential ethical issues involved in all stages of the research. We cannot just be data-gatherers".

7.10   ASSISTIVE TECHNOLOGY RESEARCH—A SUMMARY

Assistive technology aid should, as far as possible:

- satisfy the expectations of potential users;

- be well engineered;

- be well marketed (effectively but without too much hype); and

- be supported by proper infrastructure of after sales service.

The research and development should involve clinicians and potential users at all stages of the design and evaluation—but they should not dominate the development.

There are some methodological challenges which are unique to the field of AAC research, but do provide useful reminders for other assistive technology research. An effective communication aid should supplement the user's other communication methods (gesture, facial expression, etc.), rather than replace them completely. AAC research has to take into account:

- What language skills does the client have?

- Do they have normal, poorly developed, or no written language skills?

- Do they have other language skills such as Bliss?

- Do they only have the ability to communicate in pictures such as Makaton?

- Should the device be personalizable for individual users? and

- The output of the device should respond to the expectations of potential listeners as well as potential users.

There are many reasons for the failure of AAC systems, and these are technical as well as non-technical. A major non-technical one is insufficient training. This can apply to all assistive technology, but a particular problem with AAC devices can be that the user does not understand conversation and its rules. Having the right tool does not mean that automatically one becomes a skilled craftsman. Personalization is usually necessary for AAC devices, but designers need to be careful that they do not use this excuse, and the excuse of the need to provide a very flexible system, to abdicate their responsibility for designing for actual users, and for making the system effective and efficient for all potential users.

> *Encourage serendipity.*

The main challenge is developing good ideas, and every effort needs to be made to encourage serendipity within the design team, and to ensure that there is an intellectual infrastructure that forms the bedrock for the development of new ideas.

> *"I can call up spirits from the dead…But will they come when you do call for them"?*
> **Shakespeare. Henry IV, ACT I Sc 3**

Particularly in assistive technology research there is a need for really radical ideas, and I would argue that, unless there is significant opposition to an idea it cannot be radical enough to have the potential to make substantial improvement to our field (and many others). Researchers need to remember that radical research is seldom without mistakes and false starts—but they should learn from them rather than bury them (Chapter 2 gives examples of success based on failure).

CHAPTER 8

IT Systems for Older People

Information Technology has great potential to improve the quality of life of older people, but much of it has not been designed with this section of the population in mind. This chapter describes e-mail and web browsing systems designed especially for older people, together with a range of systems to assist in daily living, and to reduce social isolation and provide entertainment for people with dementia.

8.1 INTRODUCTION

Alm et al. [2002] discussed the advantages to older people of information technology in a paper to the International Conference on Universal Design in Japan: "Older people and Information Technology are Ideal Partners". They pointed out that Communication and Information technology (CIT) has the potential to play a major role in assisting older people to take part more fully in life. It can transform very small physical movements into powerful effects, and can amplify sensory and cognitive abilities. CIT has transformed the human activities of communication, entertainment, and shopping, but it has primarily been younger people who have benefited. Keeping in regular touch with friends and family, shopping from home, and accessing high quality entertainment are all goals which older people are likely to want to achieve. When effective ways are found to make these services easily and acceptably available to older people, large new markets will become available.

- Communication and social connectivity has a major role to play as loneliness and social isolation are increasing problems faced by many older people and their families.

- Being able to shop and access services from home is clearly an advantage for older people with impaired mobility.

- Remote access to telecare and telemedicine could help to reduce the costs of medical care and play a part in encouraging a self-help approach to keeping healthy.

- CIT offers the possibility of remaining economically active and productive through opening up new information—handling job opportunities and allowing people to work from home.

  However, this is all dependent on CIT becoming attractive to, and easy to use by, older people.

Figure 8.1: Elderly people are an important and growing market segment.

8.2   CYBRARIAN: AN E-MAIL SYSTEM DESIGNED FOR OLDER PEOPLE

Office systems, such as word processing and e-mail designed for "expert users" in the workplace, have been installed in home computers without any thought of the different user characteristics, requirements, or support services of home versus office software. As an alternative approach, the UK Department for Education and Skills (DfES) funded a project to develop the "Cybrarian" system. This was to be designed to "facilitate access to the Internet and to learning opportunities for those who currently do not, or cannot use the Internet because of a lack of skills or of confidence or because of physical/cognitive disabilities". The system was to be "attractive to older users (over 60 years of age) who were uninitiated and unconfident in the use of computers and for whom the internet was an alien territory" (DfES Briefing Document, 2003).

The Dundee "UTOPIA" team acted as advisors to a software consultancy house (Fujitsu) who were commissioned to produce a "proof of concept" design for a web browser and e-mail system [Newell et al., 2007b]. Fujitsu were very experienced engineers, who had significant user-centered design experience, and were fully aware of the accessibility guidelines of the W3C consortium. However, they had not previously designed systems for older people. On the basis of our research, including in-depth interviews with older users of e-mail systems, we made a number of suggestions. We pointed out that a serious challenge novice users face is that existing interfaces are overly complicated and contain far too many options (see also Hawthorn, D. [2002]). This results in a very crowded screen with small targets and menus. Mike Smith (in Dickinson et al. [2005]) calculated that the standard "out of the box" version of Microsoft Outlook Express at that time contained one hundred and one possible alternatives directly and a further one hundred and twenty nine indirectly. This "functionality bloat" is very widespread: as an example, in Figure 8.2 an "out of the box" version of Yahoo is shown. Experienced users hardly notice this complexity, but this provides an example of what causes the comments of older people reported by Morris, J. [1992]: e.g., "Having so much information presented at once is mind-boggling."

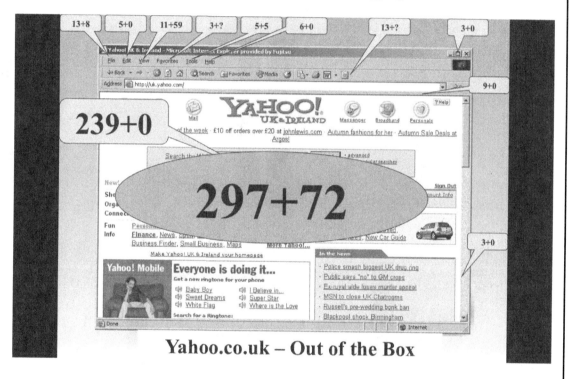

Figure 8.2: Designed for older novice users? (thanks to Mike Smith, see Newell et al. [2007b]).

The "proof of concept" design was aimed to produce uncomplicated interfaces, combined with display characteristics suitable for older adults with minor visual impairments which addressed issues of background knowledge, and burden on memory. We also believed that it was important for users to "know where they were", and to remove opportunities for error as much as possible, or at least for the system to be "forgiving" of them. A set of guidelines for the designers were produced, which Dickinson et al. [2005] lists in full.

Paper prototypes of this interface were developed via several iterations interspersed with heuristic evaluations and walk-throughs with older users, and a workshop with plenary sessions and small breakout groups where participants interacted with the paper prototypes. The value of using paper prototypes in this phase of the development was that the engineers were not as committed to these ideas, as they may have been if they had had to invest time in software development.

The final system, which was fully implemented, allowed users to read, send, reply to, delete, and forward e-mails. Each screen had a single primary purpose with only one text entry box. Figures 8.3 shows the "compose" and "read" message screens, and the web portal of Cybrarian. The default text was 14 point font with a minimum target size of 32 point. Instructions were provided on each page in non-jargon form to reduce memory demands. Accessibility options were available via a "personalize" button. All actions were text entry or a single button click.

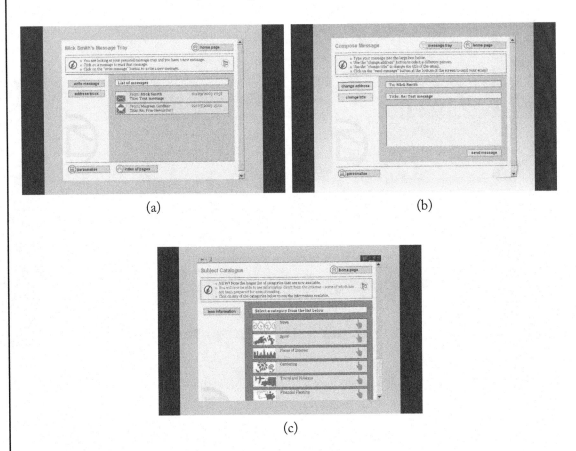

(a)

(b)

(c)

Figure 8.3: A web portal for older people. (a) The Cybrarian e-mail system—message screen; (b) the Cybrarian e-mail system—compose screen; (c) the Cybrarian web portal.

Fifteen people aged between 58 and 87 evaluated the system and compared it with Outlook Express. Each user attended two, one and a half hour sessions over a two-week period. The evaluations indicated that Cybrarian was easier to use, those people using it needed significantly less help and made significantly fewer errors and hesitations than when using the comparison system. The participants described it as easier to use, more comfortable and easier to remember than the comparison system.

This project formed the basis for the MyGuide web portal

(www.Myguide.gov.uk)

The UK government made this available to over 3500 local UK on-line Centres. In 2011 their Business Support Team reported that: "[MyGuide] has been a fantastic tool for the UK online centres network and in the last seven years has helped more then 500,000 people take their first steps online". To have taken a group of older people with little knowledge of computers and a perception of the Internet as 'alien' and received very positive responses to the experimental system, coupled with demonstrations of successful use in a large scale real situation, shows that it is possible to develop appropriate systems for those people who are currently unlikely to use the Internet.

Many of the characteristics of the current Cybrarian system are also appropriate for other users and potential users and for other IT systems. As well as being helpful for novice users, the methods used in this project are likely to be useful for those with mild sensory cognitive or motor control impairments—or even people who are just too busy/tired/drunk to wish to grapple with the complexity offered by most software interfaces.

Other research into interfaces for older adults point to the need for systems with low functionality. For example: Carroll and Carrithers [1984] addressed the problem of younger novice users becoming lost in menu options. By stripping down the available functionality, and making many confusing options inaccessible, they considerably assisted learning.

8.3   SENIORMAIL: ALSO DESIGNED FOR OLDER PEOPLE

Using a similar approach, Hawthorn [2002] developed SeniorMail, an e-mail system for older users based on Microsoft's Outlook Express but with a much simplified menu system. It was different in detail to the Cybrarian system, but had similar characteristics. His design brief included:

- avoid complex drag and drop type manipulation;

- reduce the complexity of menus;

- reduce the need to scroll;

- remove all unnecessary features;

- provide linear navigation;

- make it simple to learn;

- not require knowledge of how Graphical User Interfaces work; and

- provide confirmation feedback wherever possible.

Hawthorn had substantial user involvement in his work. He began by observational studies of older people attempting to learn to use computers, together with discussions in focus groups. He followed this by conducting three case studies involving older people in both the design and evaluation process. He found older people learned more slowly than younger people, had difficulties in finding features on the screen, suffered navigation problems, and only slowly learnt from the errors they made. Hawthorne found a number of "perpetual novices" who could use IT systems in a very basic way but seemed to be unable to progress.

All participants in the evaluation were surprised and pleased by their success when using Hawthorne's systems. He concludes that design for older people should not proceed from isolated facts about ageing, but a number of aspects should be considered in combination, including large fonts, together with simple linear predictable search spaces, limited functionality and minimal scrolling. Hawthorn's PhD thesis [2006] provides a wealth of qualitative data on older people, the relationship between their physical, sensory, motor and cognitive abilities, and the challenge of interacting with computers, and also their psychological reactions to computer technology. These data, and the insights he obtained, more than justified the time he spent on deep personal engagement with older people and in observational studies. He does comment, however, that the research literature includes so many suggestions that they are likely to overload a designer. Hawthorn, D. [2000, 2003] also developed interactive tutorials called WinTutor, and FileTutor that helped to highlight the problems his older users were having.

8.4   SUSTAINING THE USE OF CIT

Not only do CIT systems need to be designed to encourage older people to start to use them, it is also important to design them so that old people do not give up using them through frustration, or as their ability to use such systems declines. SusIT,(Sustaining IT use by older people to promote autonomy and independence) is a multi-university project, led by Leela Damodaren at Loughborough University [Olphert et al., 2009]. The project has been examining the socio-technical challenges of the digital divide, and investigating both social and technological solutions to these challenges. They have developed a user group of over 300 older people across the East Midlands. In addition to collecting survey data these older people are seen as an active part of the research strategy. Their research has found that barriers to digital engagement include:

- being scared of "breaking it" and lack of help when the technology "does go wrong";

- a lack of awareness and information; and

- a lack of effective and affordable learning opportunities.

Some participants made particularly scathing comments about computer courses they had attended.
A particularly worrying result from their survey found that approximately 30% of the survey respondents had given up using at least one piece of digital technology, a quarter of whom because it was too hard to remember how to use it. This confirms the comment in the 2004 report of the

UK Digital Inclusion Panel that "there is risk in the medium to long term, that significantly more citizens will migrate from being digitally engaged to being unengaged, rather than the other way round, as their capabilities change". It is clear that it is not sufficient to persuade older people to start to use digital technology, there is also a significant challenge in sustaining their use of this technology.

The SusIT project suggests a framework for improving the sustainability of digital technology with older people by combining socio-technical theory [Cherns, A., 1976] with participatory/inclusive design and management of change and knowledge sharing techniques. Important aspects considered should include: ethical questions, security and trust, and an understanding of the changing nature of the home environment. They suggested that users' socio-technical needs could be met by a combination of novel and adaptive interfaces designed for older people, and social networks providing access to information and support both in the home and the community.

8.5   ASSISTIVE TECHNOLOGY FOR OLDER USERS

There has been a large amount of research and development of specialist devices for older people. At Dundee these included the development of smart house technology, to which John Arnott and Nick Hine contributed important work examining the requirements for smart houses [Morgan et al., 2008]. McKenna and Nait [2005] examined the use of video scene analysis software for detecting falls within a smart house. Both these groups used theatre techniques to facilitate focus groups of older people and care workers as part of the requirements gathering exercises (see Chapter 12).

Various projects have examined the development of memory aids and Inglis et al. [2002], together with Ann Wilson of Cambridge University developed a system based on Personal Digital Organizers. A significant part of this research involved working with clinicians, psychologists and people with memory impairments to develop a robust set of requirements before developing the technology.

Mark Rice addressed the issues of developing a home telecommunication system via digital television for older people, and compared a chat system, a scrapbook of memories and reminder system based on digital television technologies. His research led to a number of innovative interfaces for application of this type [Rice et al., 2007].

8.6   IT SUPPORTING DEMENTIA

Developing IT-based systems for older people provides major challenges, but my colleague, Norman Alm, decided to tackle what at first sight seemed to be an impossible task—developing IT systems which can be used and enjoyed by people with dementia. Dementia can involve the complete loss of short-term memory which can rule out most social activities and interactions. Conversations with people with dementia can prove to be very difficult and distressing for friends and relatives. Although short-term memory may be very seriously impaired, many people with dementia retain long-term memories [Rau, M., 1993], but have difficulties in accessing these memories within a conversational setting. Reminiscence therapy attempts to access long-term memories by providing people with dementia with prompts, such as photographs, music, or artifacts from when they were young. Reminiscence sessions are known to be valuable, but are not carried out as frequently as would be desirable because of the preparation time necessary—searching for and providing material in an appropriate form can be very time consuming.

8.6.1 CIRCA: A MULTI-MEDIA SCRAP BOOK

With the aim of facilitating conversations with people with dementia, Alm, along with colleagues Arlene Astell, a psychologist, and Gary Gowans, a graphic designer, decided to develop what was conceived of initially as a multi-media scrap-book [Alm et al., 2007]. They collected appropriate data such as text, photographs, videos, and songs from the past life of the city, and developed a number of prototype interfaces to access a multimedia reminiscence experience. These were demonstrated to people with dementia and their care staff at a local Day Centre. The presentation produced a great deal of interest from both staff and people with dementia. An iterative development process resulted in the development of "CIRCA", a communication support for people with dementia. CIRCA, had a hypermedia structure with reminiscence material as content. It contained photographs, music, and video clips from the years 1930–1980. Based on suggestions from the user population and their carers, there were three themes: Recreation, Entertainment, and Local Life, and the user was able to select and enjoy photographs, short music pieces and video clips related to the theme selected.

Crucially great care was taken to remove any obvious "computer look" as it was thought that this would produce a negative response. There was neither keyboard nor mouse, the reminiscence content being accessed using a touch screen with very large on-screen buttons. Figure 8.4 shows examples of CIRCA's navigation screens. Great care was taken to ensure that any sessions with CIRCA would be entirely error free—it was not possible to appear to become lost during navigation. All members of the team were exposed to people with dementia and their carers within a normal environment (a day centre) before and during the development, and evaluations were conducted informally throughout the process. The formal evaluations were carefully constructed so as not to distress the person with dementia.

The prototype was evaluated both in one-to-one and group settings with people across the range of dementia severity, and the effects were compared with sessions involving other enjoyable activities. Traditional reminiscence sessions usually involve a series of questions from the carer followed by one response from the person with dementia. In contrast, the CIRCA sessions were more of a conversation, with each person contributing an equal amount, and control of the direction of the conversation being shared. The system elicited entirely positive reactions from care staff and relatives. People with mild to moderate dementia were able to make full use of the system. More severely affected people could not actually engage with changing the system display, but showed a marked reaction to any musical items that appeared. CIRCA is now commercially available, and the team are investigating other applications for people with dementia, including interactive games which can be enjoyed without the need for working (short term) memory.

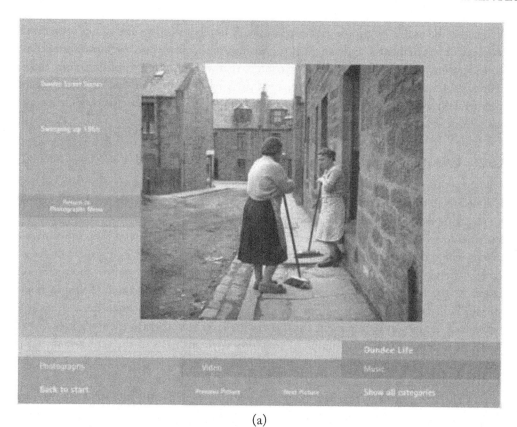

(a)

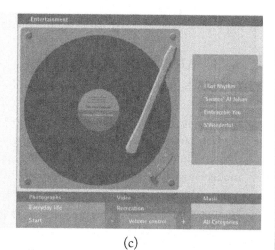

(b)   (c)

Figure 8.4: (a) CIRCA—a multi-media reminiscence therapy tool; (b) CIRCA in public house; (c) CIRCA's record player.

"Living in the Moment" was also developed by Alm's team. This is a set of interactive games designed to be usable by people with dementia. A number of ideas for games were tried out with people who had dementia, and the most promising approaches were developed into full prototypes. Consideration had to be given to devising content that was appropriate and engaging, and prompting methods to compensate for the users' impaired working memory. The first set of these games is now commercially available. Work continues on expanding the number of effective and enjoyable game types [Alm et al., 2009].

8.6.2   CREATING MUSIC

Following the success of this approach Riley and Alm developed a system called "Express Play" designed to allow people with dementia to compose and perform their own music. This had a simple but engaging touch screen interface. Screen shots of Express Play are shown in Figure 8.5. The user is invited to select an instrument and a mood (happy, sad or angry), and then move to one of three screens where they can play either happy, sad, or angry music by dragging their finger around the screen. The user hears chords playing whilst also seeing a circle appear on the screen under their finger. As the user's finger moves, so they begin to draw on the screen, leaving a trail of circles behind. This provides a continuous prompt that something happens when the screen is touched—a particularly important requirement for those with severe short-term memory loss. Users can play chords of different pitch and volume by touching different parts of the screen and may be able to develop a tune. A formal evaluation established that the system engaged the users, and they created novel music that was particular to the individual user [Riley et al., 2009].

These developments clearly show that carefully designed software and hardware combinations can be operated by people with severe cogitative dysfunction, but such developments require teams with the appropriate mix of expertise and close links with potential users and their carers so that an appropriate and effective list of requirements can be developed (see Chapter 11).

8.6.3   AIDS FOR DAILY LIVING

In a collaboration between the Universities of Dundee and Toronto (Canada), Hoey et al. [2010] has been investigating how leading edge CIT systems can be used to assist people with dementia to complete basic activities of daily living. The particular application area chosen is hand washing. A video camera records the hand-washing activity, and image analysis software is used to track the person's hands and towel and categorize the particular behavior. A probability and decision theoretic model is used to determine the state of the action and determines whether and what prompts to offer to the person (e.g., if the person has not used the soap but is moving to drying their hands, then an appropriate verbal or audio visual prompt will be given). As a fail-safe mechanism, the system can call for the assistance of a human carer. This system was tested in a Toronto long-term care facility by six older adults with an average age of 86 years—five having moderate and one severe dementia. An overall decrease of 25% in the need for caregiver assistance was found with some users no longer needing any assistance. Mihailidis et al. [2008] give more details of this system. The group

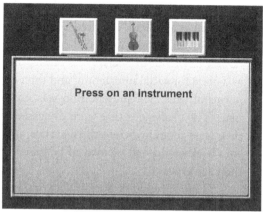

(a)

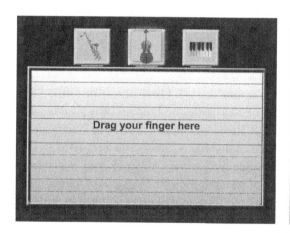
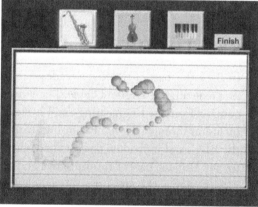

(b)

Figure 8.5: Express Play music making software for people with dementia. (a) Start screen and choice of instrument; (b) making music on a touch screen.

are now extending their research to other activities of daily living such as mobility, tooth-brushing, and cooking. The key to the uptake of such technology has been found to be the ability for the end-users and caregivers to customize the systems to suit their individual needs.

8.7   SOCIAL NETWORKING FOR OLDER ADULTS

Lehtinen et al. [2010] reported an interview study of four men and four women, between the ages of 58 and 66, relating to social networking in Finland. They reported that these respondents did not see any point in Social Networking Sites, deeming them useless. They did not see the Internet as a place for social interaction, and their perceptions of social relationship were incompatible with norms of sites. Self-presentation was not considered socially acceptable and privacy was also an issue. The challenges that were discussed were the participants' lack of confidence in their skills, their fears about malicious third parties, and of making accidental social blunders. They rejected the sites because they all seemed to be based on social norms of young people, rather than people of their age groups.

This rejection of Social Networking Sites is particularly sad as, in principle, such sites could alleviate the social isolation of many older people. Thus, research is needed into how such sites could be improved. Lehtinen et al. [2010] suggested that such sites could be improved if they were designed to provide more of a "club atmosphere". These authors also suggested that this research could have implications for all age groups making the point that "older adults may be the ones who still dare to question the usefulness of technology".

8.8   THE DIGITAL DIVIDE

In the latter part of the first decade of the 21st Century, the UK Government prioritised research into reducing the Digital Divide. The Universities of Newcastle and Dundee were awarded a £12M contract from the Engineering and Physical Sciences Research Council in 2010 to investigate "Social inclusion through Digital Economy" (SiDE). Their industrial collaborators were IBM, NCR, and the BBC, and individual projects had additional academic partners. Vicki Hanson was responsible for the Dundee part of this initiative and a range of research projects focussed on transforming the lives of people who are currently excluded were initiated. The pool of users that had been developed at Dundee will be expanded to 3000 users in Scotland and the North of England with a wide variety of ages, technical skills and abilities.

These projects include:

- Cognitive modelling (in collaboration with IBM and Carnegie Mellon University), will characterise the behaviour of computer users experiencing age-related decline. These models will be embedded into tools to assist designers.

- SEEDS (in collaboration with the Universities of Kent and Leeds) will obtain a better understanding of needs of older people, how to educate designers, and how to incorporate these data into the design process.

- Access Framework (in collaboration with BBC and IBM) will develop an Open Source tool to mitigate effects of functionality bloat in computer systems, by automatically making system corrections which the users can either accept or reject. This framework will also serve as a general platform for accessibility and cognitive support software.

- Adaptive Interfaces will be developed to enable applications to proactively provide customised interactions. Initially this will be focussed on people with low vision and mobility needs with indoor navigation and digital T.V.

- MemoJog will investigate the use of reminders on mobile phones. Carers will be able to remotely schedule and check reminders, and emergency messages will be automatically sent if there is no response to important reminders.

- Social Networking sites for older adults will be investigated to ascertain the design issues and privacy concerns which are inhibiting older adults from benefitting from this technology as a way to understand why the digital divide exists.

- The Portrait project will involve working with families of people with dementia, to create a software tool which will provide a multi-media portrait containing personal and social information including key life events, family preferences, hobbies and interests. This will be designed to enable care staff to get to know a person with dementia quickly.

- The Give Blood project will use social media and smart phone technology to create a more engaging and participative experience for blood donors.

- The New Technologies to support Older People at Home by Maximising Personal and Social Interaction project will be conducted in collaboration with the (Engineering and Physical Sciences supported) Rural Digital Economy Hub located in Aberdeen University. It will create healthcare solutions for rural environments, and achieve a better understanding of the interactions between care and social contact with reference to healthcare technologies in the home.

8.9 CONCLUSIONS

In order to increase the amount of IT that is older people friendly, designers and policy makers need to be aware of the market opportunities created by the increasing numbers of older people in the population. Designers also need to understand:

- the challenges IT provides to older people;

- older people's physical, sensory, and cognitive characteristics;

- the motivations, needs, and wants of older people;

- methods for developing appropriate designs; and

- effective ways of interacting with this user group.

CHAPTER 9

Designing IT Systems for Older People

By the beginning of the 21st century governments and other agencies were becoming aware of the increasing numbers of older people in the world, and the challenges this implies. CIT has an important role to play in responding to many of these challenges. It is thus important for designers and policy makers to be aware of the demographic challenges, and also to provide data on the characteristics of older people, and insights on their motivations, and behaviors to assist CIT designers in their task.

9.1 OLDER PEOPLE AND DEMOGRAPHICS

In 1994, I gave a keynote speech at the Inaugural symposium of the Japanese Chapter of ACM [Newell, A., 1994]. Although this was focussed on people with disabilities, I noted that by the year 2000 10% of the population of the developed world would be over 80 years old. I thus suggested that mainstream engineers should be concerned with the challenges presented by older, as well as disabled, people. Sandhu and Wood [1990] reported that in 1987 there were 29 million people over 65 in the European Union, 14% were over 80, and these were the fastest growing sectors. My comments, however, were particularly relevant to Japan which had the fastest growing proportion of older people in the world.

As early as 1980, and prior to the home computer revolution, specialists warned that, if the potential of technology to support older people was to be realized, their needs should be recognized and considered [Danowski and Sacks, 1980, p.128]. The failure of the software industry to heed this warning means that we are now faced with a situation in which standard computer technology is inappropriate for a large, and growing, proportion of the population.

In 2003, the UK Department for Education and Skills (DfES) estimated that there were as many as 24 million people in the UK who did not or could not use the Internet. Of these, 8.8 million adults in Britain did not use the Internet because of lack of confidence or skill, and 4.2 million felt they were "too old" to use the Internet. The use of computers and the Internet were found to be very skewed towards younger people with educational qualifications. Few un-skilled people over the age of 40, and the vast majority of people over 65, did not use the Internet. There is also considerable evidence from other sources that older people distrust modern technology [Jessome and Parks, 2001], and feel uncomfortable with it [Gilligan et al., 1998, Williamson et al., 1997]. This distrust and discomfort normally translates into reluctance to use many modern technologies. A survey carried out by Universities of Glasgow and Dundee confirms that the older a person is, the less likely they are to use recently developed technologies [Goodman et al., 2003].

9.2   OLDER PEOPLE'S USE OF INFORMATION TECHNOLOGY

This situation has not changed dramatically. Roberts [2010] reports that older people's use of the Internet remains lower than that of younger adults, 64% of those over 65 never having used the Internet. He also reports that, although the usage by older people is growing, it is doing so at a slow rate. In her "Manifesto for a Networked Nation", published in the UK in 2010, Martha Lane Fox reported that one in two people over 65 years of age did not use the internet, nor did three quarters of those over 75. The situation in the U.S. is not dissimilar. Czaja and Lee [2008] reported the 2005 figures that only 26% of people over 65 years of age used the internet compared with 80% of those between 30 and 45, and they were less likely to use other technologies such as auto tellers and VCRs.

Instilling (or at least not reducing) older people's confidence in being able to use CIT is a major design challenge, particularly as the very notion of what is "old age" is being transformed. Far from being a time of universal decline, Cohen, G. [2001] argued that the weakening of physical faculties is compensated for by an increase in holistic judgment, a lifetime of experience, and a release of creative abilities and energies, due to reduced work and family responsibilities. Current and emerging technology has great potential in assisting older people to participate more in social and economic life, and of considerably enhancing older people's productivity, and their communication, and support systems.

The stereotype of an older person is one with reduced abilities, and thus suited to the assistive technologies that are available for disabled people. Designing for older people, however, is not necessarily the same as the traditional design of CIT for people with disabilities. The traditional "disabled user" of assistive technology products is relatively young, usually having a single disability, and a high motivation to use such technologies. Syme et al. [2003] also found that "accessibility options" designed for disabled people can lack clarity about the effect that they will have and this can promote insecurity and confusion particularly in older people. She reported an example of "high contrast" settings with titles such as "eggplant" (black on green) and "rainy day" (black on blue). There are also problems with previewing accessibility settings - some are implemented immediately, others when the user has clicked 'apply' and closed the dialog box, and a third group are only implemented when the computer has been reset. In addition a significant proportion of accessibility research and development has focussed on blindness, even though this is not a particularly prevalent disability. In the same way as the wheelchair is the symbol of a disabled person in the "real" world, it seems that a blind person has become the symbolic disabled person in the virtual world. Thus, the characteristics of "disabled people" for which much CIT based assistive technology has been designed are significantly different to those of older people.

I returned to the theme of designing for older people in the early years of the 21st Century. At a keynote in the 2002 ASSETS Conference (published in Newell and Gregor [2002]), I asked the question "design for older and disabled people—where do we go from here"? I also reported the start of the UTOPIA (Usable Technology for Older People: Inclusive and Appropriate) project in Scotland [Dickinson et al., 2002] which was led by Dundee University and involved three other Scottish Universities (Glasgow, Napier and Abertay Dundee). This project was established to examine the issues surrounding older people's use (or non-use) of technology [Eisma et al., 2003]. Its aims were:

- to raise awareness in CIT designers of the opportunities and challenges of an ageing population;

- to change attitudes of mind to what older people need and want from IT;

- to provide a framework for checking accessibility, effectively and efficiently; and

- to introduce a revised design pragmatism sensitive to the needs and wants of older people which would lead to systems which are neither patronizing to older users, nor totally unsuitable for the rest of the population.

9.3   MAINSTREAM INFORMATION TECHNOLOGY AND OLDER USERS

Older people have as much, if not more, to gain from traditional CIT than the rest of the population, but much mainstream CIT seems to have been designed by, and for, a young male computer whizz-kid who is obsessed by technology, and who likes to play with it to discover all the things that it can do. Much of the rest of the population learn to live with this as they either see the need for, or cannot function without, this technology. They can just about manage to use it—despite frequently getting very angry. Many older people, however, do not have the resources to cope. Hawthorn, D. [2002] and Dickinson et al. [2003] report the serious problems that older novice users face with existing interfaces that are overly complicated and contain too many options. This complexity results in older users becoming confused and/or lost in a host of alternatives competing for their attention. Their difficulties are often exacerbated by multiple minor impairments in vision, fine motor control and reduced cognitive functioning, including attention span, and short term memory.

Czaja and Lee [2008] reported that older people are worse than younger people in many computer tasks including difficulties in:

- reading text on the screen;

- selecting targets, and perceiving tool bars and icons;

- hearing auditory prompts;

- clicking and pointing and dragging; and

- cognitive tasks: attention processes, working memory, discourse comprehension, problem solving and reasoning, inference formation, and interpretation and encoding and retrieval in memory [Park, D., 1992].

Sutcliffe, A. [2002] made a similar point about multi-media. He commented that "currently design leaves a lot to be desired" and needs to "adopt a usability engineering approach". Many other studies have shown that older adults have more difficulties in search and retrieval and were found to use less efficient navigation strategies [Czaja and Lee, 2008]. Freudenthal, D. [1997] suggested that deep menu structures may not be appropriate, as navigation is dependent on special skills which tend to reduce with age. Although older people are receptive to use of computers, they report more anxiety, less self-efficacy and less comfort. Nair et al. [2005], and Fisk et al. [2004] underlined the importance of the usability of the system, both hardware and software.

Many older people find the Windows environment and many web pages and applications very confusing and difficult to use. In the 2010 UK OFCOM survey, 60% of those older people who were recorded as not using the internet said that it was "not for me" and 30% said " I do not have the skills". It is probable that many of these were too proud to say that they are so scared and/or confused by the technology as not to be prepared even to try to use it. This effect is not confined to traditional information technology. Mobile telephones also require good vision and a high level of dexterity, and are becoming increasingly complex. They are not very popular with the older population, and many of those who have them only use them for emergencies.

In our own research, older people have made comments about computer and web technology such as:

- It's "totally bemusing", "very complex and very difficult", "overpowering".

- "There is too much information, it is very, very confusing and irritating".

- "(When I use the computer) it goes anywhere it wants rather than where I want it to go".

- "I find sometimes I suddenly lose everything and [find it] difficult to get back. Sometimes I get something I did not want and do not know how I got it".

- "I will have forgotten what to do in 24 hours".

- "I am not sure why people want to go on line".

9.4 THE CHARACTERISTICS OF OLDER PEOPLE RELEVANT TO THEIR USE OF IT

There is an urgent need to review the design of mainstream information technology so that it is more usable by older people. Czaja and Lee [2008], however, commented that "there is little empirical knowledge to guide these developments and almost no research has been done with older adults—especially important given demands on cognitive processing, visual search, working memory &

selective attention". There are also major issues concerned with older people's lack of confidence with technology and the effects of negative stereotyping on this phenomenon.

There is a wealth of research data about the various age-related impairments that have an effect on people's abilities to use technology (see, for example, Carmichael, A. [1999], Dickinson et al. [2003], Kline and Schieber [1995], and Czaja and Lee [2008]). Disabilities associated with increasing age include reduced:

- visual acuity and contrast and color sensitivity (especially blue), and ocular motor control;

- peripheral vision and a lessened ability to disregard irrelevant parts of a display, (and an increased susceptibility to glare);

- visual search skills and ability to detect targets;

- manual dexterity;

- hearing, particularly high frequencies and difficulties in discriminating between relevant sounds and noise; and

- Cognitive functioning.

There are other important factors that relate specifically to older people:

- The rate of decline in functionality (which begins to occur at a surprising early age) can increase significantly as people move into the "older" category.

- The environments in which older people live and work can significantly change their usable functionality—e.g., the need to use a walking frame, to avoid long periods of standing, or the need to wear warm gloves.

- Language can also be very different. Today's older people come from an age where a "chip" was a piece of wood, "hardware" meant nuts and bolts and "software" did not exist.

- Computers demand much implicit knowledge—when to single-click and when to double-click (those with poor manual dexterity may not be able to double-click fast enough). The way scroll bars work (50% of our older novice users thought they should work the other way—moving the bar down should move the page down).

- It is extremely easy for the novice user to make apparently catastrophic errors, and they have no idea how to recover from them (e.g., by clicking on a window in the background, rather than on the one they are working with, the user can "lose" the window in which they were working).

- Older learners, in general, take a considerable period to attain "reasonable effectiveness" [Hawthorn, D., 2002] which includes remembering about, and learning to use, skills that 'advanced' computer users would regard as automatic.

It is, however, important to retain a sense of proportion, particularly in terms of cognitive ability and not be over influenced by stereotypes. Deary et al. [2009], of Edinburgh University, examined a "Mental" survey that measured the cognitive abilities of 11-year olds in Scotland in 1932. He re-tested a sample of these 66 years later in 1998, and found that there was little change in the rank ordering of the people tested, and much less deterioration in the top performers than he expected. (Unfortunately, many of the lower performers had not survived to be tested in 1998). It is known that fluid intelligence falls with increasing age, but crystallized knowledge remains, and there is evidence that older people use both parts of their brain when problem solving, in contrast to children who tend to only use the right side. Cohen, G. [2005] suggested that older people are better able to integrate the past with new experiences and this can lead to a new and original solution, whereas young people are more focussed. Older people can have access to much wider experiences and knowledge of the world than younger people, and a more mature approach to problem solving. This knowledge, however, is not necessarily useful for using current information technology, and, in some cases, can be a positive hindrance.

Staugh [2010] suggested that, although older people tend to forget names and sequences are prone to distraction, and cannot work out how to use new gadgets, these challenges are relatively trivial. In contrast, she claimed that on a range of cognitive skills, the middle-aged brain (40–60 years) outperforms all other age groups. Although older people are less able to retrieve information, she reported huge gains in important forms of brain performance, such as cutting through the intricacies of complex problems to discover concrete answers, and being more able to draw connections and see problems in a wider context, thus having better judgement. The (trivial) cognitive challenges of being an older person outlined by Stauch, however, are exaggerated by the interfaces of much information technology, such as their excessive functionality, seemingly random commands with no command structure and an over-emphasis on novelty.

9.5   REDUCING FUNCTIONALITY

It is essential to keep the functionality of systems to a minimum. One of the most difficult processes during the development of the Cybrarian system (described in Chapter 8) was recognizing when our preconceptions about what was "necessary" for an email system allowed complexity to creep into the interface designs. Jargon and complexity had a nasty habit of creeping into designs, and great care needed to be taken to avoid this. A very rigorous requirements analysis was required to ensure that only that functionality that was really needed was implemented. The Cybrarian development clearly showed that one of the easiest way of solving a problem is to add complexity to the interface.

The need for a very rigorous requirements analysis, and an extensive investigation of alternative solutions to problems involves a paradigm shift in managers', designers' and developers' understanding of what a good computer system is, and a recognition that a market exists for simpler interfaces. The development of a system with reduced functionality, and a very simple interface, was far from simple itself. Various design tradeoffs had to be made to retain the simplicity involving difficult design decisions for the development team.

9.6   TECHNICAL GENERATION EFFECT

Howard and Howard [1997] suggested "one reason older people have difficulties with current IT products may be because they belong to a different technology generation", and the knowledge and skills they have acquired during their lives may not be transferable to the use of current IT. Becker, H. [2000] noted that during the formative period of life (between 10 and 25 years old), people acquire values, norms, attitudes, behavior, and skills that usually stay with an individual for a long time and influence future behavior. Rama et al. [2001] confirmed this, and Hawthorn, D. [2000] commented that once a skill or information is learnt, it is harder to unlearn it, or show flexibility in using it. Older people come from a different technical generation and this impacts on their use of modern IT.

In an extensive discussion of the "Technical Generation Effect", Lim, C. [2010] conducted a semi-structured interview with six users from the "Electro-Mechanical" (direct control by push buttons, dials and switches) generation and six from the "Digital-Software" (control via menus and remote controls) generation. This revealed that the former group had more difficulties in modern operational procedures, particularly with unnecessary features and functions. This was followed by an experimental study with 35 volunteers between 18 and 83 years old using both pre-digital and digital telephones, cameras and radios. In general, older participants (aged 55 and above) in the study had more difficulties in using the layered interfaces associated with digital technology. This is consistent with findings in a web mail study by Hawthorn, D. [2003]. On the basis of this research, Lim developed a "generational time line" and used it to illustrate these effects to design students.

9.7   DESIGNING FOR DYNAMIC DIVERSITY

In very rough terms, older people can be divided into three categories:

- "fit older people"—who do not appear, nor would wish to be called, "disabled" but whose functionality, needs and wants are different to those they had when they were younger;

- "frail older people"—who have one or more "disabilities", often severe ones, but will also have a general reduction in many of their other functionalities; and

- "disabled people" who grow older - whose long term disabilities may have been affected by the ageing process, and whose ability to function may be critically dependent on their other faculties, which may also be declining.

The major difference between older people and the people for whom most assistive technology had been designed is that they have multiple minor sensory, motor and cognitive disabilities, in addition to any major disabilities they may have acquired. These minor disabilities may interact in such as way as to make the use of technology a major challenge. For example, one traditional approach to blindness is to use synthetic speech: an older person with poor hearing will find it difficult to hear the speech (particularly high frequencies) and the additional cognitive load needed to understand synthetic speech may overload their (reducing) cognitive capacity. Older people's abilities are likely to change much more rapidly than is the case with younger people, and thus a static system may no longer be appropriate. Not only is the range of the sensory, motor and cognitive characteristics found in older people much greater than those in younger people, the rate of change of these characteristics can be much greater. Figure 9.1 illustrates the changes in functionality that occurs as people age. The range of people's functionality increase with age, but also "high functioning" people, tend to retain their abilities and, even at a late age, will have greater functionality than many middle-aged people.

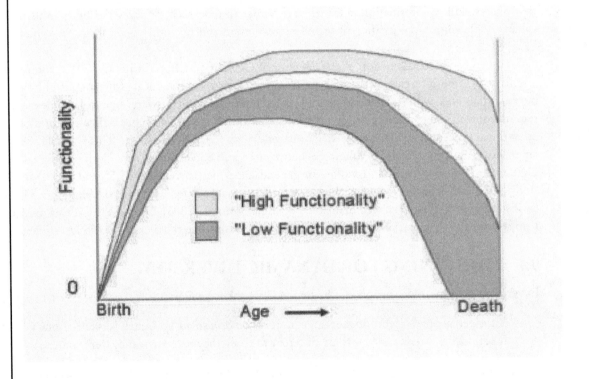

Figure 9.1: Functionality versus age.

Newell and Gregor [2002] suggested the concept of Design for Dynamic Diversity—that interfaces should be designed so that they could dynamically change as the user's characteristics changed. They suggested that that it should be possible for interfaces to change automatically, or be changed gradually, as the individual user's functionality changes. For example, as people grow older their visual capabilities reduce, and interfaces should provide an easy and progressive way of increasing font size without the necessity to make radical changes to the interfaces. Similar considerations apply to dexterity and memory capabilities of users.

Experiments with older people showed that the concept of a dynamic system allows poor memory levels and lack of confidence to be factored into the design of a product effectively and efficiently. Gregor et al. [2002] reported on the development of BrooksTalk—a web browser for visually impaired users which was designed taking into account the concept of Design for Dynamic Diversity. This particular system also had been originally developed for visually impaired users and had speech output. In an evaluation with older users 82% were unable even to get started with it, but the addition of a personal voice help facility enabled older users to use it successfully.

Sloan et al. [2010] examined the potential of adaptive interfaces, for people with age-related declines in their capabilities. He investigated ways in which adaption could be implemented in the context of web usage. Initial evaluations of ideas such as adapting font and widget size indicate that older people would appreciate the benefits of such an approach to their changing characteristics, although concern was raised at changes to the interface occurring automatically with no user involvement. Adaptive Interfaces provide a promising approach to the challenge of Dynamic Diversity, but need to be designed with care so that they do not add further to older people's confusion with technology.

9.8   CHALLENGES FOR TODAY'S YOUNG AND MIDDLE-AGED PEOPLE

It is often claimed that the challenges referred to in this chapter are only temporary, and today's young and middle-aged people will be familiar with computer technology, and thus they will not have any problems in later life. This claim rests on two assumptions: (1) that the characteristics of today's middle-aged people will not change with age, and (2) that there will be no significant changes in interface technologies. Both these assumptions are very dubious.

(1) People's sensory, motor, and cognitive skills are extremely unlikely not to decline as they age, and they will begin to find on-screen text and buttons too small, and the cognitive load of interacting with complex computer interfaces more difficult to cope with.

(2) Many of today's old people were very familiar with their generation's technology, but have not been able to cope with new technologies (In one of our focus groups, a man of 80+ said that he was a radar operator during the 1939–45 war and could take radios to pieces and reassemble them, and they would work, but he cannot cope with today's computer technology). Unless there is a significant change in the ways interfaces are designed, this challenge is very likely to be repeated with today's so-called "technologically savvy" generation. The introduction of multi-touch, speech, gesture, and other interface technologies, together with novel interaction metaphors and models could lead to significantly different ways of interacting with future products. These could be as alien to this generation of CIT users as current interfaces are to today's older generation.

The need to consider the older people's challenges with information technology products is unlikely to go away in the future. If we are not to continue to disenfranchise the older generation we need to remain vigilant. Attention paid to developing technology with a specific consideration of the needs of older people now and in the future will not only bring social and economic dividends in itself, but will help in the design of technology which is more accessible for all users.

CHAPTER 10

Ordinary and Extra-Ordinary Human Computer Interaction

Design for older and disabled people has tended to be considered a specialist area confined to Rehabilitation Engineers and Assistive Technology developers. This is one of the factors that has led to older and disabled people finding many mainstream IT systems either difficult or impossible to use. There are, however, substantial similarities between the interface needs of older and disabled people and so called "able bodied" people. The concept of "ordinary and extra-ordinary human computer interaction" was developed to encourage synergy between these disparate design groups.

10.1 REHABILITATION ENGINEERING AND HCI

Although Rehabilitation Engineering and Human Computer Interaction (or as it was originally called Man-Machine Interaction) had much in common, they were two almost entirely independent research cultures. The practitioners attended different conferences, read different books and academic journals, interacted with different academic disciplines, and were exposed to very different commercial cultures. A common belief was that the rehabilitation field benefited from "spin-off" from mainstream engineering advances, but not the other way round. This view was enshrined in the UK's Chronically Sick and Disabled Persons' Act, in the mid 1970s, where the Secretary of State was given a responsibility to report on work done in government establishments that may benefit disabled people.

Some mainstream researchers have investigated the rehabilitation field to find uses for technologies that were not good enough for mainstream use. "If it is no good for anything else what about assisting disabled people". This encourages the view that a "quick fix" is all that is required to develop assistive technology. Unfortunately, however, the problems presented by disabled people can be highly complex and often need a long-term commitment and significant dedicated research. The market place for assistive technology is also very different from, for example, military procurement, and researchers may not be aware of these differences. This approach is not entirely without value. It was essentially why I, a young researcher in Automatic Speech Recognition, was encouraged to develop VOTEM (see Chapter 1.2), and, although VOTEM did not become a commercial reality, the experience was the catalyst for my long-term interest in developing assistive technology. Also, the only commercial application of speech recognition and synthesis that has lived up to its predictions of market penetration is for disabled people ([Grandstrom, B., 1990] and [Newell, A., 1992a]). Although mainstream market penetration of these speech systems has improved, they still have not lived up to the early promise claimed for them. This shows there can be commercially viable assistive technology markets for products that do not sell well in mainstream markets.

> *Technical pull can be a useful strategy.*

Rehabilitation engineers tend to be focused on their particular user population—who they see as individuals, sometimes to the detriment of the wider ramifications of their work. Rehabilitation Engineering can also be seen as only addressing small unprofitable markets—which may imply low cost, and intellectually and technologically trivial solutions. On the other hand, mainstream designers can see design for disabled people as a fringe activity with a charitable rather than professional status. My research has led me to a very different view of rehabilitation research and its relationship to mainstream research and development. I first presented these ideas at a Speech Input/Output Conference [Newell, A., 1986], using examples from speech technology, but the message applies to a much greater range of technologies. I suggested that mainstream speech technologists could learn a great deal from considering the challenges presented to rehabilitation engineers by their disabled users, and this could lead to better HCI design practice and better human interfaces for everyone. It was also in that year, at CHI '86, that Shneiderman, B. [1986] said that "We should be aware of subtle design decisions that make use more difficult for people with physical and mental disabilities, and for individuals from different cultures…. and not create a situation where the disadvantaged become more disadvantaged".

10.2 MAINSTREAM EXPLOITATION OF ASSISTIVE TECHNOLOGY RESEARCH

I was commissioned by the Department of Trade and Industry to perform a wide ranging investigation into the relationship between rehabilitation research and main-stream HCI research, and found that the "spin-off" was very much a two way street [Newell, A., 1988b]. A number of highly successful mainstream products had been first developed as aids for disabled people. One of the most ubiquitous is the cassette tape recorder that was developed by a company making talking books for blind people—their customers having found a reel-to-reel tape recorder difficult to use. Engineers at the time said that it would never become popular because of the poorer sound quality. This is an early example of engineers underestimating the importance of ease of use.

Other assistive technology devices that found their way into mainstream markets were reported to me following a postal questionnaire and individual discussions with researchers. These included:

- long-playing records, developed for talking books for the blind in 1935;

- multi-tracked tape recorders for the same market in early 1950s;

- carpenter's mitre blocks;

- the typewriter (invented for a blind Italian Countess) and the ballpoint pen (developed for people who could not use a pen with a nib because of poor manual skills);

- early remote control systems (developed for motorically disabled people, as was the use of the electrical mains to carry control signals round a house);

- the "Concept" keyboard, (developed so that physically disabled children could use the BBC microcomputer, became very popular and is an early example of a touch screen interface to a personal computer);

- TV subtitles (for hearing-impaired people became very popular in public houses and other noisy venues with television screens);

- a large button telephone (developed for disabled people, became a top seller for the British Telcom and was also found useful in situations where people had to wear gloves (this remains a challenge with most mobile phones—either because of the size of the screen and/or it not being sensitive to a hand wearing a normal glove));

- text messaging over land lines for deaf people (pre-dated mobile telephone text messaging by many years); and

- predictive text systems used in mobile phones (were first developed for motorically disabled people).

It seemed to me that the mainstream HCI community was missing out on research challenges that could influence research. Also, if this idea was propagated in the HCI community, it could raise the profile of the important HCI challenges presented by disabled people, and thus encourage more research in this field. I thus developed a concept I called "ordinary and extra-ordinary Human Computer Interaction.

10.3 ORDINARY AND EXTRA-ORDINARY PEOPLE AND ENVIRONMENTS

I first presented these ideas at InterCHI 93 in Amsterdam in a keynote lecture entitled "CHI for everyone". (Published in Newell, A. [1993a], Newell and Cairns, [1993b, 1995], and later updated in Newell and Gregor, [1999]). The arguments I presented included that:

1. "able-bodied" and "disabled" as a descriptors for people is a false dichotomy—the characteristics of "disabled people" are usually only an exaggeration of those of the able bodied;

2. a consideration of people with disabilities underlines the importance of individual differences in HCI;

3. considering of the needs of disabled can raise awareness of the real needs of human beings;

4. research into extreme situations can tell us a great deal about human processes;

5. there are many situations where able-bodied people can be significantly handicapped by the environment in which they are situated; and

6. both groups suffer from the effects of a restricted bandwidth between the user and the equipment.

10.3.1 "ABLE-BODIED" AND "DISABLED"—A FALSE DICHOTOMY

There are situations, such as welfare payments, where it is useful/essential to define people as "able-bodied" or "disabled", but this is only applicable to a very small range of domestic and other equipment. Every human being has a set of sensory, motor, and cognitive abilities—some of them could be described as "ordinary" and some "extra-ordinary". A very substantial number of people have visual impairments, but are otherwise "able bodied". A person with good eyesight, however, may be physically weak, or have limited cognitive capabilities. These individual dysfunctions, however, are rarely different in quality from "ordinary" abilities (e.g., there are a few totally blind people—most people with visual impairment simply have poorer eyesight than a person with "perfect" vision. A mentally handicapped person has lower but not zero cognitive abilities). If we consider human beings as points in a multi-dimensional space that describes their abilities, "disabled" people do not sit in a discrete volume in that space. The situation is actually much more complex than this, because each human being moves around this space during their lives—with time constants of years (e.g., growing older), hours (becoming tired), minutes (the effects of imbibing noxious substances) and seconds (an accident).

10.3.2   INDIVIDUAL DIFFERENCES

Human Computer Interface research often seems to assume that "the user" will be an "ordinary" person with "average" abilities. Some research may divide users into "novices" or "experts", but it is often assumed there is little or no variability within these groups. This approach is obviously not suited to the rehabilitation field, but I would argue is also not appropriate for mainstream HCI. A greater awareness of individual differences in user characteristics and behavior would benefit many mainstream designers.

10.3.3   A CONCENTRATION ON REAL NEEDS

It can be extremely difficult for severely disabled people to control equipment, and designers need to be sure what controls they really want and what is superfluous. The "functionality bloat" of much office software shows very little evidence that the designers of such systems have ever attempted such an exercise, even though the usability of such software would benefit greatly from the removal of unnecessary functionality.

10.3.4 LEARNING FROM EXTREME SITUATIONS

One of the major ways we learn about the human body is from medical research that examines the ways in which it malfunctions. Automobile manufacturers invest in the development of racing cars partially on the basis that designing for extreme conditions will lead to design improvements in their commercial products. This paradigm of learning from extreme cases could play a major part in HCI research. Examining how interfaces perform when the user has motor, sensory, or cognitive disabilities can provide useful ideas as to how to improve its performance for all users, and for "able-bodied" users suffering from fatigue or stress.

10.3.5 HANDICAPPING ENVIRONMENTS

People can be handicapped by their environments. Smoke produces similar effects to visual impairment: high levels of noise, hearing impairment: the need to wear gloves, reduced tactile sensitivity and dexterity: extreme surfaces, such as sand–reduced mobility. Fatigue and stress can also cause significant reductions in cognitive abilities. Thus, a soldier on a battlefield will be deafened by gun-fire, blinded by gun-smoke, physically impaired by body armor, and cognitively impaired due to being scared of being shot! In HCI terms this person is severely disabled (I asked the audience at a workshop I gave to a space agency group to describe the user for whom they were designing equipment. They started to describe a person at the height of physical and cognitive ability but, after due consideration of the effects of the space suit, they decided that within 15 minutes their users were more dead than alive!). These are examples of how an ordinary (able-bodied) person in an extra-ordinary situation has similar characteristics to an extra-ordinary (disabled) person in ordinary situations. Less extreme parallels exist in many non-military situations. These include "hands busy" situations, and interfaces designed for use by a one-handed person can be appropriate when the user has to stand, and there is nowhere to rest the equipment. The major application of automatic speech recognition—despite 50 years of research—is for disabled people and people who cannot (or it is difficult for them to) use their hands for tasks that they have to perform.

 The human interfaces to equipment can be very disabling, and can benefit from interface options developed for disabled people. An excellent example of this in practice is the inclusion of predictive text options within mobile phones as described in Chapter 4.

10.3.6 HUMAN COMPUTER INTERFACE "BANDWIDTH"

The information between the human and the computer is transmitted via a channel whose bandwidth is limited by the users' abilities and the HCI interface. This restricts the amount of information that can be passed between the user and the machine, and can become overloaded. For example, the pilots of a combat aircraft may not have the visual capacity to receive as much information as desirable, and their dexterity restricts the amount of control they can have of the aircraft. This situation is only quantitatively different from a person who, because of visual and/or motor impairment, cannot use a word processor effectively. In both these cases, we are seeing the effects of a reduced bandwidth, and the technical solutions to these apparently very different challenges are more similar than might appear at first sight.

10.4   PITFALLS OF NOT CONSIDERING THE NEEDS OF PEOPLE WITH DISABILITIES

There are thus important motivations for considering the HCI needs of older and disabled people in addition to the drivers of demographic trends, compliance with disability legislation, and social pressures. There also can be unforeseen downsides if this is not done. Many blind people lost their jobs when the Windows environment replaced the command line interfaces of business systems. Accessibility options for this new interface lagged many years behind the introduction of these systems. There is a danger of history being repeated with the growing popularity of touch screens, which, although being potentially a better interface for some disabilities, provide significant challenges to others. Apple launched the iPhone without support for two key accessibility needs—text-to-speech and a global minimum font size option, although a number of accessibility features have been added since the original model was released in 2007. Some very successful screen readers have been developed, but there is (in 2011) little to help people with large fingers or impaired dexterity.

From an academic perspective, the danger of ignoring the rehabilitation field is that one is unaware of important and relevant discoveries. This was seen in the CHI 1986 conference where a paper on a "foot mouse" showed total ignorance of the fact that disabled people had been operating equipment with their feet for many years, and the authors of a paper on "auto-completion in full text entry transcription" seemed to be unaware of the previous work in this area.

10.5   ECHO AND ARCHIE—INSTANTIATIONS OF THE ORDINARY AND EXTRA-ORDINARY CONCEPTS

In the early 1990's, Cairns et al. [1993] was developing a multi-modal workstation for people with disabilities—specifically for a physically impaired non-vocal telephonist. The system was called "ECHO" and included speech, eye-gaze and data-glove input, and visual and synthetic speech output. It also included rate enhancement techniques that had been developed in other projects.

This project was expanded to respond to the ideas of ordinary and extra-ordinary HCI in a collaborative project with Computer Resources International (Denmark), GEC-Marconi Avionics, Bertin & CIE (France), Sofreavia, and UK Civil Aviation Authority [Ricketts et al., 1995]. The concept was to build a system called "ARCHI" which would act both as an office workstation for a disabled person (an extra-ordinary person in an ordinary situation) and as a cockpit aid for the pilot of a high performance aircraft (an ordinary person in an extra-ordinary situation). The project incorporated multi-modal presentation and control together with methods of monitoring, inferencing and prediction from our disabled work with the intelligent knowledge based planning and prediction systems from the avionics field. It used real-time user modeling to produce an adaptive and personalized system that would form the basis of an efficient workstation for a disabled user and improve pilots' performance by increasing the effectiveness with which they could control their aircraft. It would also provide a system that the pilot could use if they were injured in combat. The results from this research fed into our own future projects, as well as providing useful data and design ideas for our avionics colleagues.

10.6 TAKE-HOME MESSAGES

Figure 10.1 illustrates how the different design philosophies map onto user populations. Rehabilitation engineering and accessibility options focus on those with below average function-

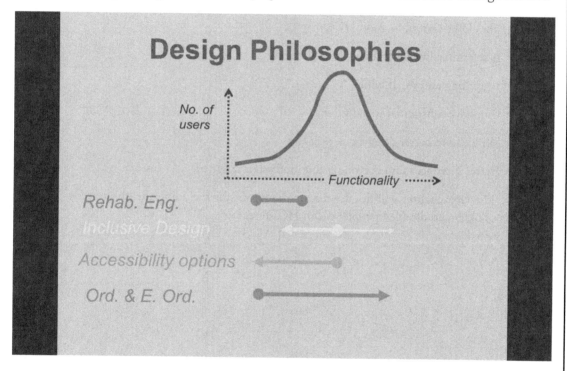

Figure 10.1: Design philosophies.

ality. Inclusive Design and Accessibility options tend to take products designed for average users and extend their user base to include people with below average functionality. The concept behind "ordinary and extra ordinary HCI is to focus the design on those with below average functionality with a view to developing systems which encompass the needs of people with both average and above average abilities.

Designing for people with below average functionality should be an important part of the tool kits of all designers. The reasons why researchers need to be aware of rehabilitation and assistive technology research include:

- most people have ordinary and extra-ordinary abilities;

- most people will grow old and/or become seriously disabled;

- extra-ordinary needs are only an exaggeration of ordinary needs;

- everyone can be handicapped by their environment; and

- solutions for extra-ordinary people can produce better interfaces for everyone.

Research focused on extra-ordinary (older and disabled) people:

- Should be central to good HCI,

- Is scientifically and technically challenging,

- Is exciting and worthwhile,

- Can have substantial social value,

- Will lead to better design practice, and

- Better interfaces for everyone—able-bodied or disabled.

The next chapter will discuss research methods for including the characteristics, needs and wants of older and disabled people within HCI research.

CHAPTER 11

User Sensitive Inclusive Design

Current methodologies for human computer interaction and information technology design describe and list a vast amount of useful information. Designers will find these very useful. They should be conversant with this literature, and also with those methodologies and guidelines focussed on the inclusion of older and disabled people in the potential user group for products.

This chapter addresses the challenges of working with extra-ordinary people in the design process. It focuses on the insights which have been gained from the author's and his colleagues' research in this field. The chapter will suggest concepts that should be considered by designers when they embark on a project involve this user group rather than exhaustively listing methods and guidelines.

11.1 UNIVERSAL DESIGN/DESIGN FOR ALL

Few designers positively set out to discriminate against older and/or disabled people—they may have simply been seduced by the stereotypes that: less able people do not provide a good market, that such considerations will "ruin their designs" and make them less "cool" and saleable in mainstream markets, or they may not believe the costs of such exercises justify any possible benefits. Some developers have even complained about "rewarding inexperienced un-skilful thick people" and that "stupid people skew results"—quoted in Knight and Jefsioutine [2002]. The lack of uptake of usability recommendations is also attributed to communication failures between usability experts and designers [Blythe, G., 2002]. Designers and commercial decision makers often overestimate the costs and under-estimate the market advantages that such an approach brings. In order to increase designer's focus on older and disabled people they need to be persuaded that this is a good idea.

"We are the music makers. We are the dreamers of dreams".
Arthur O'shaughnessy (late 19th cent.)

In the middle to late 1990s a number of groups began to address these issues. Vanderheiden, G. [2000] gave a well-received keynote lecture in which he provided much data relevant to this debate and described many examples of the advantages of mainstream designers considering older and disabled people within their potential user group. Shneiderman, B. [2000] published a paper discussing "Universal Usability" in the Communications of the ACM, and in the preface to Lazar, J. [2007]

provided a useful history of Universal Usability. This approach suggests that designers should ensure that all potential users, including older and disabled ones, be considered within the design process.

Other names have been used such as "design for all", "inclusive design", and "accessible design". Examples of such initiatives are the INCLUDE project within the European Union (http://www.stakes.fi/include).
This produced a methodology for "Inclusive Design" for telecommunication terminals [Hypponen, H., 1999] that was based on user centered design and usability principles and an extension of the International Standard for human centered design [ISO13407, 1999]. Their approach was to suggest design compromises and the use of "guidelines as a good cheap basis for integrating needs of people with varying abilities into design at an early phase". The Trace Center in Madison, Wisconsin is focussed on technology and disability and has produced a wide range of publications in this field that can be found at
http://www.trace.wisc.edu.
The Centre for Universal Design at North Carolina State University provides information and specialist courses in the more general field of Universal Design, and their publications can be found at
http://www.design.ncsu.edu/cud/ud/ud.html.
These are very similar to general user centered design principles, with a philosophy based on the premise of "Equitable Use"—that is: "the design should be useful and marketable to any group of users".

Keates and Clarkson [2004] gave an introduction to inclusive design and in Keates and Clarkson [2003] provided an extensive and detailed review of inclusive design mainly from a product design perspective. They suggested that a user's sensory, motion and cognitive capabilities can be represented within an "inclusive design cube", from which the percentage of the population that any product excludes can be calculated.

An extensive list of publications relevant to Universal Usability can be found in the review monograph by Meiselwitz et al. [2009].

"Design for all"—a realistic goal?

These movements have been extremely valuable in raising the profile of disabled users of products, and have laid down important principles and guidelines. Many "design-for-all" guidelines are excellent, but the term itself has some inherent dangers. In an everyday understanding of the words, "design for all" is very rarely achievable, and this can have the effect of less informed designers simply paying lip-service to an unachievable goal, with the message interpreted as "if you are designing a new product, take account of the needs of older and disabled users, but only if this is easily achieved". This may discourage designers from making important first steps towards a more inclusive design philosophy, and provide a barrier to greatly improved access for most. More seriously "everyone" is not a particularly useful design brief. A good designer needs to have a clear concept of the potential product and the potential user groups—"everyone" provides far too much variability to be of much use to the designer. Providing access to people with certain types of disability can also make a product more difficult for people without disabilities to use and often impossible for people with a different type of disability. The use of guidelines and authoring tools may also become an excuse not to attempt to really understand the challenges older and disabled people have with technology and replace well thought out design decisions with a "tick box" approach to accessibility.

> *"Everyone" is a poor design brief.*

An additional concern is that the impression is sometimes given that "universal design" is focussed on modifying mainstream products—recommending that, somewhere in the design cycle, designers should take account of the unusual demands of marginalized people such as older users. This suggests applying the "universal design" concept towards the end of the design cycle, which can lead to the requirements of marginalized groups being considered as an "add-on" extra to an otherwise well designed product. Not only does this patronize older and disabled people, but is also likely to lead to significantly increased costs, and possibly to inappropriate compromises which are bad for both traditional and marginalized groups of users.

11.2   THE "ACCESSIBILITY" APPROACH

An approach that can be particularly appropriate to software is the provision of "accessibility options" for people with disabilities. IBM provides a comprehensive set of accessibility guidelines in the form of checklists for developers:

`http://www.ibm.com/accessibility/us/en/.`

The World Wide Web Consortium (W3C) has also produced guidelines aimed at raising the accessibility of web content to disabled people

`http://www.w3.org/WAI.`

A detailed discussion of accessibility options can be found in Sloan et al. [2010], and Sloan, D. [2009] addressed the issues of web accessibility research.

Such accessibility options are absolutely invaluable for disabled users who, without such options, would be completely unable to use much business and leisure software, or to obtain useful information from web pages. The danger of assuming that "disabled users" will need accessibility options is that such an approach can encourage mainstream systems developers to ignore the difficulties some users may face. The perceived dichotomy between "able" and "disabled" does not reflect reality—a continuum of functionality exists between the two extremes and many users, who would find more accessible designs useful, do not regard themselves as 'disabled' (see Chapter 10 and Newell and Gregor [1997]).

Some designers also seem to focus entirely on ensuring that a disabled user is able to "access" the information (e.g., via speech output for a blind person), but do not consider the usability of the system via the accessibility option they have provided. There have been many examples of web pages in particular being fully compliant with accessibility guidelines but being completely unusable. Milne et al. [2005] found that designers need more than just W3C guidelines if the web sites they develop are to be usable by, as well as "accessible to", older people. This view was confirmed in Colwell and Petrie [1999] who, in a survey of the accessibility of websites [Petrie and Hamilton, 2004], found that the observance of W3C guidelines did not necessarily lead to a site that was usable by disabled people.

11.3 USER-CENTRED AND PARTICIPATORY DESIGN

The design community has been aware of the importance of focusing on users' needs, wants and capabilities for many years. "User-Centred Design" (UCD) is a popular approach to this challenge. This is essentially a set of techniques which encourage developers to focus on the users within the design process (e.g., Preece, J. [1994], Shneiderman, B. [1992], and Helander et al. [1997]). The British Standard (ISO 1999) states that "(User Centered Design) is characterized by the active involvement of end users and that the multidisciplinary team should include end users". Participatory Design [Muller, M., 2002] proposes a focus on the users, and includes both the pragmatic approach of direct collaboration between designers and users and a more conceptual approach that incorporates complementary perspectives to help designers come up with better solutions. Koskinen et al. [2003] have suggested an "empathetic design" model of involving users in product design. These design techniques all recommend close interaction between designers and users. Traditional User Centered Design methods, however, provide little or no guidance on how to include older and disabled people in the process.

Ethnography in another popular way of involving users in the design process [Blomberg et al., 2002]. The ethnographer acts as an intermediary between the users (whom they observe usually in their own environment) and the designers. The ethnographer's role is to present as full a picture as possible of the user, their culture and their environment to the designers. These are often in the form of a written report, but "experience models", "opportunity maps", profiles, scenarios, mock-ups and prototypes, are also used.

11.4 BUILDING ON DESIGN FOR ALL AND USER CENTERED DESIGN

The "design-for-all" approaches have been particularly effective in raising awareness of the need to consider older and disabled people, and many designers have found the concepts, standards and guidelines to be of significant practical value. They, however, do have certain drawbacks, particularly for ab-initio innovative projects that include older people in their potential user population.

In addition to providing designers with standards, guidelines and "how to do it" texts, it is important to develop in them an empathy for the user population, rather than a narrow focus on "rules". This view was developed during the Cybrarian project discussed in Chapter 8. The software engineers working on this project approached the design from a "user centered approach", but were unfamiliar with the challenges and constraints of working with a user group of older people. They were, however, very much aware of accessibility guidelines, and were keen to apply them. Ironically their commitment to traditional UCD and accessibility guidelines made it difficult to persuade them that this approach was less than adequate when designing for older people. The Dundee group provided a one-day training course focussed on the needs of older people. This included data about the characteristics of older people and an opportunity to talk with some older users. This exposure seemed to have little effect, and the developers felt that the academics were exaggerating the difficulties older people had in using computers.

It was not until the developers were persuaded to interact closely with older people, and personally observed older people trying to use paper prototypes, that they fully realized older peoples' technological ignorance, their fear of and approach to technology. The developers' views changed, and changed significantly over the period of this project [Newell et al., 2007b]. Arranging for the developers to interact with older people developed a level of empathy that we do not believe would have occurred if focus groups and evaluations had been conducted solely by usability professionals, who reported their findings to the development team. This study underlined the importance of "changing the attitudes of mind of designers", by providing them with information in a form which made maximum impact. In Petrie and Hamilton's [2004] audit of web sites, it was clear that the designers had ensured that the web pages could be read out by a speech synthesizer, but it was difficult to believe that the designers had ever themselves listened to examples of such audio output.

These results underline the need for including within the design process a method whereby the clients and developers are exposed to users characteristics in a way that has the maximum impact on their designs. Although standards and guidelines and research results are essential, it is not clear that these alone have sufficient impact. The work also showed the impact of designers actually observing older users interacting with their systems—as opposed to simply reading the results of such evaluations.

11.5   USER SENSITIVE INCLUSIVE DESIGN

The above experience led us to develop an extension of User-Centred Design that we believed captured the extra requirements needing to be considered when working with older and/or disabled people. We called this "User Sensitive Inclusive Design".

Know your users but do not rely on them especially in conceptual stages.

The use of the term "inclusive" rather than "universal" reflects the view that "inclusivity" is a more achievable, and in many situations, appropriate goal than "universal design" or "design for all". "Sensitive" replaces "centered" to indicate that it is rarely possible to design a product that is truly accessible by all potential users. The term "sensitive" also implies a different relationship with the users than the term "centered". It suggests that the users are firstly people and that the designer should develop an empathetic relationship with them, rather than treat them as "subjects" for usability experiments. This concept is discussed in detail in Newell, A. [2008].

11.6   ENGAGING WITH USERS

A major plank of the User Sensitive Inclusive Design message is to encourage designers to develop an empathy with older and disabled users through meeting with them in appropriate situations. Dickinson et al. [2002] examined various ways of involving users in a User Sensitive Design Process. This needs to be done with care with both developers and older people sensitized to what is required of them. It is important to:

- control the introduction of inexperienced developers to older users with little computer experience;

- create a situation in which the older users feel comfortable, (e.g., not a laboratory full of computer equipment); and

- underline the fact that they are not just "experimental subjects" but that their comments and insights are very valuable.

The research group in Dundee does not use traditional "usability" laboratories where the subject and the researcher are separated by two-way mirrors. We thought that such a barrier would be counter productive and made a positive decision not to include this type of usability laboratory within a new human interface research facility [Newell et al., 2006]. Our researchers are always in the same room with the users to provide psychological support and help where necessary [Newell et al., 2007a]. This resonates with Wixon's [2003] comment that "it is no accident that most usability testing involves encouraging entire design teams to watch the test, and it is well known that much of the effectiveness of the test comes from this active participation". It would not be appropriate for all the design team to sit in with a user, but perhaps a combination of a researcher sitting in with the user plus the rest of the design team watching through a two-way mirror would be a suitable compromise.

11.7   MUTUAL INSPIRATION BETWEEN RESEARCHERS AND USERS

Where possible, it can be advantageous for users to meet in small groups. We found that small groups of people evaluating paper prototypes within interactive workshops, can create a more mutually supportive environment than individuals faced with an unfamiliar design. The process of "thinking aloud" is considerably less artificial as a discursive exploration of an interface with others, than a process of laying one's uncertainties open to an "expert". We encourage the users to consider themselves as part of the design team and Eisma et al. [2002] developed the concept of "Mutual Inspiration", where we try to obtain original ideas from the older people rather than simply seeing them as a resource to check out our own ideas and concepts [Dickinson et al., 2003, Eisma et al., 2004, Newell et al., 2007a]. Gheeraw and Lebbon [2002] described a similar process which they called "empathic bonding" to stimulate creative thinking and user-facilitated innovation.

Figure 11.1: Mutual inspiration.

Although students are rewarded by a small payment, this is not always appropriate for older and disabled participants as such payments may conflict with their state benefit rules, and so travel allowances are more appropriate. Other rewards should also be considered. We suggest that any meetings that involve older users contain a significant time period for socializing between themselves and the researchers (such as at a coffee break between formal experiments, or focus group meetings). We have also found that older people are keen to hear how their input has affected the project, and suggest providing informal lectures to users at the conclusion of projects.

One valuable asset developed during the UTOPIA project was a cohort of older people who work with us. Contact was made with a diverse range of older people and a database of over

160 individuals and 24 groups was developed, which has subsequently been expanded in other projects. There was a focus on ensuring diversity in aspects such as: demographics (age, gender, class), experience with technology (computer users, novices), and inclusion of specific groups (individuals who have specific difficulties, for example mobility, speech problems). The cohort is thus diverse including people from many different backgrounds with various life experiences, ranging from IT literate people living at home to people in day centers who have never used a computer.

11.8   THE "USER CENTRE"—A SOCIAL SPACE FOR OLDER USERS

As part of a new building we dedicated a specific area as a "User Centre" [Newell et al., 2006]. This space, shown in Figure 11.2, forms a small "club room" for the older users. Although we do not give formal instruction, we provide "taster" sessions for new technologies (e.g., digital cameras, book making software, etc.), and provide opportunities for peer-to-peer learning which is particularly effective for this age group. In this way we have developed a cohort of older people who are very supportive of our work and understand the research paradigm.

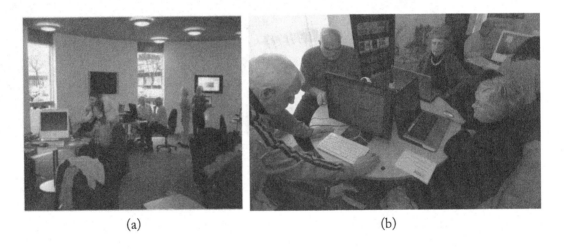

(a)   (b)

Figure 11.2: Scenes from the User Centre in the Queen Mother Building.

We have found that there are other very fruitful ways of ensuring that designers fully understand the needs and aspirations of the user groups. The long-term arrangement with a cerebral palsied non-speaking user described in Chapter 7 has meant that he fully understands our needs as researchers, and we are able to support him in non-research aspects of his life. This has led to a relationship where there is obviously a great deal of mutual support and respect. We have also been fortunate to employ a number of cerebral palsied engineers, thus ensuring that there is no concept of "them" and "us" between designers and users.

Norman Alm and others have found it particularly valuable to meet with older and disabled users within their own environments, and provide some service to them, rather than seeing them as simply "potential users". Norman spent much time with a group who were developing an internet café from scratch in a small rural village. He gave them technical and emotional support and in return gained great insights into their relationship with technology that would never have appeared in the laboratory. Our very successful work developing computer systems to support people with dementia (see Chapter 8) was in no small measure due to Norman spending a half a day a week for many weeks as a helper at a day centre for people with dementia. This was repeated by his student, Philipa Riley, who also acted as an unpaid helper at the day centre. This was aimed at familiarizing her with the lives of people with dementia, and, as it did not include any work directly related to the project, did not need the extensive ethical clearance that formal experiments with people with dementia would require—that would come later in the research project. These sessions sensitized the researchers to the true needs of people with dementia. They did not have a focus on technical support but proved immensely valuable in the development process.

We conducted in-home research via an aurally presented questionnaire. In this case, the researcher was prepared to assist with any technical problems that the person might have as well as conducting the interview. One episode underlines the dangers of a less involved approach. Whilst the researcher was setting up a video recorder (a challenge for many older people) the lady said that she often sent emails via the television. The first formal question to be asked was "do you use the internet" and the lady replied, very forcefully "no I never use the internet". Park and Schwarz [2000] have shown age differences in the ways in which people respond in self-reports, older people tending to be more cautious in their responses. This group also tends to use the "don't know" option more than younger respondents, and have even been known to pencil in a "don't know" column on questionnaires [Eisma et al., 2004].

In her highly successful DBA "Design Challenge", organized yearly by the Helen Hamlyn Institute at the Royal College of Art, Julia Cassim works with young designers and connects them with real people with particular disabilities [Cassim, J., 2005]. This is designed to show how interaction with disabled people can be a direct benefit to mainstream product and service. Each year five groups of designers are given a design brief and introduced to an expert user with an appropriate disability. They have 24 hours to develop an inclusive concept inspired by the needs of their individual user. Observers at this event have commented that this has encouraged designers to go outside the "designer bubble" and listen carefully to other people's needs.

11.9   FORMAL EXPERIMENTS WITH OLDER AND/OR DISABLED USERS

There is clearly a need for formal user groups as part of the product development cycle, but specific challenges occur when older people and people with disabilities are part of such groups [Eisma et al., 2003]. These include:

- there is likely be a much greater variety of user characteristics and functionality, and finding and recruiting "representative users" is a challenge;

- it may be difficult to specify exactly the characteristics and functionality of members of the user group;

- the functionalities of older people can change rapidly with time, and this may not be obvious from single meetings with focus groups;

- it may be difficult to get informed consent from some users, and some users may be "incompetent" in a legal sense;

- older people may tire easily thus constraining the time scales for formal experiments;

- there may be conflicts between users with different disabilities and conflicts between accessibility for disabled users and ease of use for less disabled people (a ubiquitous instance being that floor texture can assist blind people but may cause problems for wheelchair users, and people with prams or "pull-along" suitcases);

- it can be difficult to obtain accurate data. Older people tend to be very positive about the prototypes that are presented to them—wishing to praise the developers rather than give an objective view. If they cannot cope with technology, they tend to blame themselves rather than poor design. (As an attempt to counteract this bias, we tend not to say that we have designed any software we are evaluating with older users);

- many older people have a fundamental distrust and a very limited understanding of the underlying computer concepts, and are unlikely to have technological imagination (imagine discussing the motor car within an exclusively horse culture). Thus, one must know one's users, but not rely on them, particularly in the conceptual stages of a project; and

- there may be large cultural and language differences between older people young researchers which cause poor communication.

All the precautions mentioned above are important, not only for ethical reasons, but also provide more effective ways of eliciting genuinely useful information.

11.10 SUBJECT EXPERTS AND CLINICIANS

In the case of designing for disabled users, it is important to get as much information from subject experts such as clinicians. However, one should not necessarily rely on their judgments when the use of technology is involved.

> *A good subject expert is worth their weight in gold, but do not slavishly rely on all subject experts.*

Clinicians and teachers may have their own "hang-ups" about technology, and there may be conflicts of interest between their needs and those of their clients. It is also important to draw the distinction between needs and wants. When considering older and disabled people, clinicians tend to focus on "needs" whereas in their personal life most people are more concerned with what they want rather than what they need. There is also the assumption that older and disabled people want "independence".

Needs or wants—independence or control?

Such rhetoric forgets that increasing wealth is almost always used to reduce independence. The British upper classes employed "footmen" essentially as a replacement for their feet (so they did not need to walk), and the rich employ chauffeurs and gardeners, but the "master" is always in control. The challenge is to ensure that older and disabled people are supported by assistive technology, but still retain control of their lives. Many older people are isolated and lonely, and information technology should be used to encourage and increase contact with people rather than replace it.

11.11 A FOCUS ON AESTHETICS

There is a tendency to assume that older and disabled people have little or no aesthetic sense and, unlike other user groups, are motivated entirely by the functionality of products. Hocking, C. [1999] reports that in the U.S. 56% assistive technology is quickly abandoned, and 15% are never used. Waller et al. [2005] suggested that a lack of user centered design may contribute to the poor adoption of AAC technology and that the inclusion of end users in the design process may reduce this level of abandonment.

Another factor in this very high level of abandonment of assistive technology products may be poor aesthetics. There can be an assumption that, when designing for older and disabled people, novel and beautiful concepts need not be considered, the design team being focussed exclusively on the ergonomic and technical aspects of the product. With the major exception of eyeglasses, assistive technology is mainly designed either to be invisible (e.g., hearing aids) or as medical products more suited to a hospital environment than a domestic one. This was not always the case: in Victorian times a walking stick was considered a fashion accessory, using a stick being a badge of honor rather than shame, and many very beautiful walking sticks were produced. More recently walking sticks clearly indicate that the user is either disabled or a hill walker. Except for a small number of relatively attractive decorated folding walking sticks, function has totally over-ridden form. Pullin, G. [2007] suggests that assistive technology can be designed to be attractive, and this could significantly increase the usage and retention rate of these technologies, as well as improving the self-perception of users.

Technology should delight the user as well as assisting them.

In his book "Design and Disability", Pullin, G. [2009] discussed in depth the use of techniques more often found in creative design houses such as IDEO (London) (see also Pullin and Newell [2007]). One such technique is Critical Design, which is used in creative design houses mainly for product design [Myerson, J., 2004], but could have a role in the design of assistive technology systems. Critical Design is used to ask carefully crafted questions about the future path of technical development, rather than directly providing solutions. Dunne, A. [1999] argued for the use of Critical Design techniques in designing for older and disabled users. Disability groups and researchers are already beginning to exploit critical design as a tool to provoke discussion about issues that may otherwise go un-discussed, such as "should hearing aids be invisible". In 2005 the Royal National Institute for the Deaf exhibited the "Hearwear" project at the Victoria and Albert Museum [Thompson, H., 2005]. This included a range of beautiful hearing aids designed to be conspicuous, and other communication aids, such as one designed for use in noisy Public Houses, which could be of value to hearing as well as hearing impaired people.

Pullin and Cook [2010] addressed the role of prosody in Augmentative and Alternative Communication aids using critical design techniques. They developed a range of "speaking chairs" each of which instantiates a mental model of how prosody could be included into AAC devices. Chairs are used as they support the idea of face-to-face communication and also enable prototype electronic systems to be hidden. They were designed to represent the principles of an interface rather than a particular design outcome. They provide a memorable experience for the users/audience and facilitate discussions of what is the most appropriate way for users to control the tone of voice of speech synthesizers.

11.12 A FOCUS ON EXTRAORDINARY USERS

User-Centered Design often suggests that a small group of people should be recruited who can be considered to "represent" the proposed user group for the project, but this can be extremely difficult given the large range of user characteristics of older and disabled people. It is not easy to be certain that one's sample of users is "representative" even with a very constrained user group, but virtually impossible to produce a small set of older users who are truly representative of this population. Pullin and Newell [2007] have suggested that design teams should examine the characteristics of the product and choose a number of "extra-ordinary" users who could be considered to be outliers rather than representative users. Considering this group as a number of individual people rather than "examples of disabilities" facilitates the development of an empathetic relationship with potential users. Only after considering a number of individual design solutions tailored to each "extreme" user, does the team attempt to bring together the various designs to produce a final design, or numbers of designs. They suggest that this approach is likely to inspire more radical solutions, and lead to designs that are truly innovative without sacrificing their accessibility. Choosing and interacting with users in this way can also provide richness to the design process and encourage the emergence of radical solutions. Dunne and Raby [2001] made the point that "populations can validate a design, but individuals can inspire new thinking and therefore are invaluable at the beginning of a project".

Design teams, which ideally should include industrial designers, interaction designers human factors specialists and engineers working together, should be encouraged to consider a number of specific "outriders" in depth and, initially at least, design for them in particular. More comprehensive designs can follow at later iterations in the design process.

> *Dream about what is possible and desirable rather than what is feasible (current technology) and what needs to be done (the currently problem).*
> **[Holnagel, E., private communication]**

Design teams should focus on what users really want rather than what clinicians and other experts say they need. It is valuable for a combination of techniques taught in Art Schools, human-computer studies, and more traditional engineering design methodologies to be employed in the design process.

CHAPTER 12

The Use of Professional Theatre

A major recommendation of user-centred design is to work closely with users. Interactions between designers and older and disabled people, however, can raise practical and ethical issues, and often the communication between designers and user groups is not as effective as it could be. This chapter suggests the use of theatre professionals as mediators between these two groups. A number of projects are described in which live and filmed professional theatre was used both to raise designers' awareness of issues and as part of a requirements gathering exercises.

12.1 AWARENESS RAISING

Successful design, which includes older and disabled people in the potential user group, requires designers both to achieve an empathy with their potential users, and have access to sufficient relevant human factors knowledge about their intended end-users' needs, wants and abilities. Although a large corpus of information exists about the abilities and requirements of older people, much of this is either inaccessible to, or not used by, designers. Designers also need more 'soft' data about users, such as their problems, preferences, lifestyles, and aspirations, and this can be particularly important at the earliest, conceptual stages of design.

The designers in the "Cybrarian" project (described in Chapter 8) did not really understand the issues older users had with technology until they actually observed them using systems and discussed the issues directly with them. Within educational or commercial development setting, however, this is not always feasible. There are organizational challenges in contacting and bringing together user groups, and interacting with older people requires significant skills that some designers may not have. The question is thus: "Is it possible to improve designers' empathy with older users without them necessarily having to have one-to-one contact with them?"

Methods which have been used include "personas"—fictional characters instantiating an array of qualitative data representing the user group [Cooper, A., 1999] and scenarios illustrating how users may react to equipment within realistic situations [Benyon and Macaulay, 2002]. These methods are being taken up with enthusiasm by parts of the design community, including software and interaction designers [Head, A., 2003, Pruitt and Grudin, 2003]. It is claimed that personas promote engagement of designers over a period of time, which in turn promotes insight into users' goals and the ways in which users might respond to design features. Grudin [2006], however, commented that "the (portrayal of the user) is not engaging…it is not generative—it provides no handle for thinking about a new situation". Also, the emotional attitudes of the users, that can be as important as their physical and sensory abilities, tend not to feature in traditional personas. An alternative approach is to use ethnographers as intermediaries between users and designers.

We were looking for powerful communication tools aimed at designers with little or no experience of inclusive design, and postulated that professional theatre could be very effective in transmitting important messages about user characteristics to designers. A range of theatrical techniques have been used by other researchers. These include a documentary approach, actors performing various specified tasks with the technology, and designers themselves acting out various scenarios in front of their peers. The use of actors in design development was reported by Salvador and Howells [1998], and Sato and Salvador [1999]. Dishman, E. [2003] and others used actors in unscripted live drama which Dishman calls "Informance design".

> *The play's the thing, Wherein I'll catch the conscience of the King.*
> **Shakespeare, Hamlet, Act II Scene 1**

We wished to encourage dialogue within design communities, and between designers and users, as a way of changing the mind sets of designers. We thus needed a theatrical genre that was specifically designed to encourage audience participation. We studied the ideas of "Forum Theatre", developed by Boal, A. [1995], and further developed by the Foxtrot Theatre in Education Company (Dundee, Scotland). This company initially, and more recently M.M. Training (Dundee, Scotland), used these techniques extensively within professional training for communication skills (e.g., within palliative care, and the training of medical students) and in community consultation (including with seniors). The companies worked closely with the School of Computing to develop a form of theatre that was particularly appropriate for our application. The director of both companies, Maggie Morgan, was the Leverhulme Artist in Residence of the School in 2005/6. We also included a studio theatre, designed specifically for interactive theatre, as part of the new building for the School of Computing (see Figure 12.1).

12.2   USING PROFESSIONALS

We use a narrative rather than a documentary style of theatre and combine this with audience interaction in the Forum Theatre style. The use of professional scriptwriters, actors, directors, and film makers ensure that the quality of the theatre is high. We commission a scriptwriter who conducts detailed research in the subject area in collaboration with the design team—gathering data and relevant anecdotes. The writer then weaves these into human-interest stories, and produces a series of short pieces that address the important research issues needing to be discussed. These pieces have a narrative style and include the emotional content and tension essential to good drama. This process itself can create conflict. The researchers tend to focus on the information they want transmitted, whereas the writer needs to produce well-structured and tight scripts with good characterization. This is a learning experience for the researchers. Via a number of iterations of draft scripts between

Figure 12.1: The Studio Theatre in Queen Mother Building.

the writer and the design team, well-constructed stories which transmit the required information and raise the appropriate questions are produced.

> *Do not despise anecdotes. Data informs—stories illuminate.*

These plays are either performed live by professional actors, or made into films. Following the performance, a trained facilitator encourages the audience to address the issues raised in the plays. If live actors are present at the performance, "hot seating" is also used. The actors remain in their role as characters in the play and reply to questions from the audience. Essentially, this is a form of improvised role-play, but done by professional role players, who have been well briefed about the situations in which they find themselves.

> *"Anecdotes are essentially faithful to the truth, precisely because they are fictional, invented detail by detail, until they fit a certain person exactly".*
> **Ernest Sabato**

Using this methodology, we produced films and presented live theatre both in educational settings and at international conferences. We have also pioneered a combination of film and live theatre, where a film is shown followed by one or more actors from the film appearing and being "hot seated" by the audience.

12.3   THE UTOPIA TRILOGY

The first set of films, the "UTOPIA Trilogy", addressed the challenges older people have in using technology, and were dramatizations of some of the issues the researchers had encountered during the UTOPIA project. Based on real events, conversations and observations, they were the amalgamation of many and were designed to convey older people's experiences with technology and the situations they encounter in a thought-provoking, but humorous way.

The stories used were:

"Peter and Jane Buy a Web Cam".

"Sandy's Mobile Adventure".

"Email Experience".

Brief descriptions of these films are shown in Figure 12.2 and the UTOPIA trilogy films can be found at:

http://www.computing.dundee.ac.uk/projects/UTOPIA/.

One advantage of using theatre is that scriptwriters and actors know when to exaggerate for effect, and how to articulate feeling in such a way that it communicates effectively to the audience. This was illustrated in the "Email Experience", one of the characters lost his temper with the technology and the air "turned blue". This was thought to be out-of-character by one of the researchers

PETER AND JANE BUY A WEBCAM

Having been given her son's old computer, Jane has set about learning the basics and now feels confident using it for email and word processing. Having come across an article in a paper, she has decided to take the plunge and buy a webcam so that she can talk to her daughter and grandchildren in Australia.

SANDY'S MOBILE ADVENTURE

Sandy has been given his daughter's old mobile phone. He never uses it but carries it around with him to keep her happy. Today he finally finds a use for it when he locks himself out of his house...

A few months later...

After his problems with the phone, Sandy's daughter gave him a quick lesson and a cheat sheet of simple instructions. Sandy still doesn't use the phone, but carries it with him in the car fully charged.....just in case.

A year and a half later...

Since saving him a long walk home, Sandy has changed his attitude towards mobile phones and now uses his frequently. Over a glass of wine his daughter brings this change up...

EMAIL EXPERIENCE

Peter is quietly jealous of his wife's confidence with using a computer, but is too proud to admit it. One day he finds that she's left it on when she's gone shopping. He decides that this is the chance to give email a go in privacy...

After the debacle of trying email on his own, Peter has bitten the bullet and signed up to computer classes at the local university. He's feeling his petulant self as ever....

Frustrated with his experiences so far, Peter decides to give email one last try. The computer class is demonstrating a new cut down email application which has been designed for simplicity, clarity and ease of use...

Figure 12.2: UTOPIA trilogy.

who was used to older respondents being much more restrained in their interaction with computers. The scene, however, resonated with the (polite) older people who were shown the final videos ("that is how I felt about the computer but did not say"). This scene thus successfully communicated the reality of the poor design, the actor providing a more accurate representation of the real feelings of older people than they had been able to communicate to designers themselves.

12.4   EVALUATION OF THE TRILOGY

These videos were evaluated with a variety of audiences including academics, practitioners, software engineers, relevant groups of undergraduates, and older people. These evaluations established that the videos provided a useful channel for communication between users of technology and designers, and changed the perceptions of both students and more mature designers of IT systems and products [Carmichael et al., 2007].

We followed the UTOPIA Trilogy by two further films that were made by Soundsmove, a Scottish based film company. "Relative Confusion" was commissioned in 2007 to show the challenges digital television can provide for older people, and, in 2009, "Relatively PC" focused on the effects of the move towards a Digital Economy and the challenges older people may have in accessing this technology.

12.5   "RELATIVE CONFUSION"

Jack and Tommy decide to surprise their sister Maureen with a digital TV system, but didn't reckon with the minefield of bewildering new technology they were about to enter.
The issues illustrated by the video include:

- users' ability to learn and their memory for new control methods.

- the effects of poor eyesight and manual dexterity, and the interaction of poor eyesight and memory;

- modal errors and the effect of cognitive load;

- loss of control due to complex interaction techniques;

- the consequences of jargon; and

- the rate of learning new functionality.

Although, "Relative Confusion" focuses on the challenges older people have with installing and using digital television, the lessons apply to many groups of naïve users and a range of new technologies.

12.6   "RELATIVELY PC"

This film follows the same format as "Relative Confusion" with the same cast. In this case, Tommy, Jack and Maureen enter the digital age. Vignettes from the video illustrate the challenges of:

- "internet banking";

- why someone needs a computer;

- "making an email";

- upgrading software; and

- on-line shopping.

Newell, A. [2009] gives a full list of challenges presented in the video. These include:

- User confusion, and too much choice,

- Visual problems, and unknown metaphors,

- Reasons for technophobic, and the effects of stress, and

- Patronizing design.

Stills from this videos are shown in Figure 12.3 and the films can be viewed at:
http://www.computing.dundee.ac.uk/acprojects/iden/

12.7   USING LIVE THEATRE, FILM AND A COMBINATION OF BOTH

The videos we have produced have been positively received both by students and designers, with substantial agreement to the view that the videos had changed their attitudes to the design of computer systems. They have provoked much discussion on these issues at international conferences. Presentations included: European Conference on Accessibility to the Next Generation Networks. (BT Centre London 16th Jan. 2007), "Best Practice in ACT Design for All Teaching" (June 2008, Middlesex University, UK) and a Master Class for the Department of Computer Science, (University of York, 5th Feb 2009).

The film and live actor combination, where the film is shown followed by one or two of the characters in the film appearing in person and answering questions from the audience, has also been used in educational and conference settings. Undergraduate and Post-graduate students reported that, although the video was interesting and informative, being able to question and discuss with the 'live' characters had more impact. Presentations at HCI and AT conferences stimulated a huge amount of discussion and argument, and made the session highly memorable for the audience. With all these very different audiences, the fact that the characters were actually actors liberated

Figure 12.3: Stills from "Relatively PC". *Continues.*

Figure 12.3: *Continued.* Stills from "Relatively PC".

everyone to say what they really thought. The 'characters' were highly believable and convincing, but the audience could attack the characters knowing that the actors were not going to take their comments personally. These combined film and live theatre events were performed at the BCS HCI Conference (London UK in 2006), the Accessible Design in the Digital World Conference (the University of York UK September 2008), HCI 2010 (Dundee, Scotland September 2010), and at the Sidney Michaelson memorial lecture at the Edinburgh Science Festival (April 2010). The age range of the audience at the Edinburgh event was from 25 to over 75. 54 out of the 75 members of the audience, who responded to a paper questionnaire, agreed or strongly agreed that "the interaction with the actors increased the effectiveness of the event", and 64% reported that the event "had changed their attitudes to how computer should be designed" (some in the audience being already convinced of the message the film was portraying).

We have used live theatre at many international conferences, and, although, a formal evaluation of the impact of theatre was not appropriate on these occasions, the audience engaged fully with the drama, and the question and answer session with the audience was very animated. A measure of the success of this approach is that, on the basis of conference presentations, we have been commissioned by other conference organizers to present at their conferences. Conferences at which we have performed include: Accessibility to Next Generation Networks, COST219ter Conference (2008 Brussels), "Access and the City", Conference, (Dublin, November 2008): the e-Inclusion Ministerial conference, (Vienna, December 2008), and two presentations at CHI 2008 in Florence, and we have been invited to perform at the 2011 conference of the Association for the Advancement of Assistive Technology in Europe.

The videos can be seen at

`www.computing.dundee.ac.uk/acprojects/iden`

CHI 2008 produced a video showing the live performance and subsequent audience participation in Morgan et al. [2008]—this gives a very authentic view of the process and can also be viewed in the ACM Digital Library:

`http://portal.acm.org/citation.cfm?doid=1358628.1358720.`

We believe that the success of this approach is due to:

- the plays being narrative based rather than having a documentary or pedagogic style. That is they illustrated the issues involved within interesting story lines, with all the characteristics of a good narrative—humor, tension; human stories, antagonists and protagonists,

- the quality of the research and the production; and

- the use of theatre and film professionals.

12.8   THEATRE IN REQUIREMENTS GATHERING

We have also used professional theatre for requirements gathering with older people. This began with a project developing the use of video analysis and artificial intelligence techniques within a system to detect falls in the home. Theatre was used to focus discussion amongst older potential users of a home monitoring system and to gain information from potential users (including sheltered housing wardens) about issues surrounding the use of current community alarms, contexts of use and what users would wish for future systems.

12.8.1   SMART HOUSE TECHNOLOGY

Following initial discussions in focus groups, four scenarios were developed and films made. These were shown to groups of older people and sheltered housing wardens. Design constraints and user requirements for a fall detection and activity monitoring system were explored. Each scenario was developed to provoke discussion of a slightly different aspect of system design. The films were produced with specific points where the action could be stopped temporarily to allow discussion. They addressed:

- general issues about falling;

- the way in which information about a fall would be sent to and received by carers;

- the information the "faller" wants and needs about the system; and

- issues related to activity monitoring rather than fall detection.

Figure 12.4 shows an example "story board" from one of these films.

The drama was found to engage the older users and focus their discussions on the functioning of fairly complex technology in an enjoyable context. These sessions were very useful in determining the design brief, and involving potential users in this way allowed them to play a real part in the design of a product. One unexpected outcome was that, in contrast to popular belief, the audience were relaxed about having video cameras in their home. We believe that this was because the films allowed them to fully realize what this would actually involve, and that it would not involve actual pictures of them being transmitted.

This project revealed a danger of presenting novel approaches in the scientific reviewing process. Our proposal included the use of actors to simulate old people falling. All the reviewers commented that this was not appropriate as actors did not know how to fall over. These reviewers were clearly not aware that a professional fight director's job is to train actors to fall over, but also the reviewers, who apparently were unaware of ethical constraints on research, seemed to be suggesting that a proper methodology would be to trip up older people!

12.8.2 DIGITAL TELEVISION APPLICATIONS FOR OLDER PEOPLE

This work was followed by using theatre for requirements gathering for a number of other projects involving older people. These included an investigation of possible applications of digital television for older people [Rice et al., 2007]. In this project short concise open-ended plays were produced illustrating:

- video-type facilities to support communication with family and friends;

- capturing and sharing memories through the television; and

- a daily reminder system.

These concepts were presented as live theatre to groups of older people, who were encouraged to discuss the issues raised. Discussion of each play lasted approximately twenty minutes, the audience members being encouraged to pose questions about the character's actions and behavior towards the technology used, whilst the actors themselves remained in character. This method proved very successful in stimulating discussion and enabling technologically naive people to understand and address important issues in relation to novel design concepts.

The audiences identified with the ideas presented, and a range of advantages and disadvantages were identified with suggestions emerging that had not previously considered by the research team. These included aspects of protocol and social etiquette, security and cost, language and terminology,

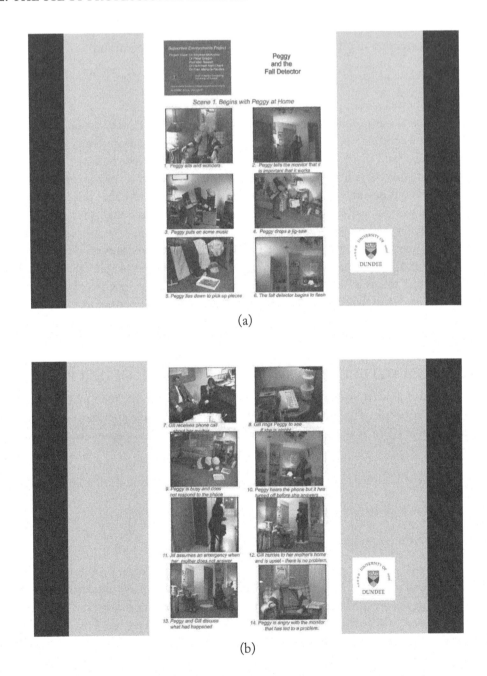

Figure 12.4: Story board of stills from fall detector project.

and the usability/accessibility of current digital receivers. A number of 'user-generated' ideas emerged, many of which became the foundation for the design and development of more detailed prototypes.

An important advantage of the use of theatre in this context was the theatric technique, well understood by audiences, of "suspension of disbelief". None of the technology presented to the audience actually worked—the systems were mocked up using various props, but the audience was not consciously aware of this. This meant that the researchers did not have to put effort into developing any equipment before they obtained user feedback. This not only saved a great deal of effort which may have been wasted, but also meant that the researchers did not have as much commitment to their ideas as they would have had if they had had to build it, and thus were more open to criticism.

12.8.3   NEW TECHNOLOGY IN THE WORK PLACE

Other situations where we have used live theatre in requirements gathering include an investigation into the effects on older workers of new technology in the workplace. Live theatre events were presented to older people at Dundee and Miami Universities and to researchers at IBM New York. Films that were based on these presentations were produced and can be seen at

www.computing.dundee.ac.uk/acprojects/iden.org.uk.

These include "Blether"—the challenges to older workers of using social networking tools in a work environment, and "A steep learning curve" that investigates different responses to computer systems and video conferencing within the workforce.

12.8.4   SMART HOUSES AND MULTIPLE STAKEHOLDER

John Arnott and his colleagues [Morgan et al., 2008] further developed the live theatre concept to conduct requirements gathering in a Smart House environment. There were multiple and disparate stakeholders, who may have had very different, and possibly contradictory interests, motivations and requirements. The audiences for the live theatre were divided into relevant stakeholder groups, who viewed the drama remotely by video link. They discussed the theatre piece within their own groups and in a plenary discussion. This process was then compared with the response of a single combined audience. In mixed sessions, the professionals tended to dominate the discussion, but when each of the peer groups had the opportunity to discuss the drama first on their own, the discussion in the following plenary session was much more balanced. The drama was found to:

- prompt a wide ranging discussion that went beyond functional utility, and included social and emotional factors; and

- facilitate a recognition of new or previously undervalued perspectives on the use of technology in domestic care, e.g., a stake-holder said "anything in use will throw up more problems that you can't predict in advance".

The researchers found that theatre assisted in the process of analyzing and understanding the needs and relationships involved in multi-stakeholder situations, and provided insights into how to develop better care systems. One of these pieces was shown in the Morgan et al. [2008] CHI presentation mentioned in Section 12.7 above.

In a further study, Arnott and his colleagues used live theatre as part of a study exploring the role of Community Matrons. Two Live Forum Theatre sessions were conducted with audiences of healthcare professionals, telecare developer and technicians. These were focused on how telecare may affect their working practices, and were designed to ask the questions of the audience:

- Are they ready to use telecare?

- Are they concerned about telecare used in self management of care?

- Would telecare data be useful to them?

- How will telecare change their jobs?

12.8.5 ADAPTIVE INTERFACES

Sloan et al. [2010] commissioned live theatre events to investigate older people's response to adaptive user interfaces. Two professional actors "interacted" with a flat panel monitor that was used as a prop on stage. An off-stage PC, operated by a member of the research team, controlled a power point presentation on the wall above the stage to simulate what the actors pretended was appearing on the on-stage prop screen. Audio output was simulated by an off-stage voice. This arrangement gave significant scope for the researchers to show a variety of adaptive interfaces without requiring any technical implementation. The audience of older people clearly understood what would be involved in an adaptive interface, and the presentation was successful in provoking discussion on the issues involved. They made comments on the pros and cons of such a system, and highlighted the characteristics of adaptive systems that they thought would be important.

12.9 THE USE OF PROFESSIONAL ACTORS TO SIMULATE USERS WITH DEMENTIA

There is also a role for actors when equipment is being developed for very frail users for whom conventional HCI experimentation may be inappropriate and/or unethical. In his research into cognitive support for daily living (described in Section 8.6.3), Hoey [2010] conduced a pilot project where he used actors to simulate people with dementia. These actors had been well briefed in the characteristics of people with dementia, including viewing many videos showing people with dementia using prototype technology. They were then used to provide training data for an adaptive prompting system. There would have been significant ethical issues in the use of people with dementia at this stage of the project, and the data gathering would have taken very much longer. This use of actors is being continued in a joint project with the University of Toronto.

12.10 THE VALUE OF THEATRE

Theatre, and particularly interactive Forum Theatre, offers new and exciting ways to inform and inspire design practice. This research taught us a great deal about professional theatre in relation to HCI research. This included that:

- scriptwriters and actors are professional observers of human behavior—in some senses they are ethnographers, but with different training;

- they are skilled in presenting that behavior in a way that engages the viewer/audience;

- their use of narrative techniques provides a human rather than technological focus;

- they know when it was appropriate to exaggerate for effect, and how to articulate feeling in such a way that it communicated effectively to the audience;

- actors are particularly skilled in "think aloud" situations;

- actors are able to present a user with a particular combination of characteristics, and could change these in response to requests from the designers (e.g., what would happen if you were older, if your sight or hearing was impaired, or if you were under pressure?);

- the use of actors removes ethical problems of "protecting the users";

- "suspension of disbelief"—the actors pretending that the technology works, means that a system can be seen being used in a "real" environment before it has been developed;

- theatre provides a natural outlet for designers and experts to engage in discussion between themselves and with (simulated and real) users; and

- theatre encourages audiences to take a creative approach to design, rather than simply presenting their views and opinions.

> *In all theatre, reality is revealed through unreality and truth is exposed by a mummer's fiction.*
> **Morris, West Daughter of Silence, Coronet 1985, p. 105**

The range of research described here has shown that professional interactive theatre can be a powerful tool within the I.C.T. development process. Our research has shown that professional actors, particularly those with improvisation skills, have the ability to:

- respond appropriately to audience comments;

- think aloud and question their own actions, whilst still remaining in role;

- communicate emotional responses to audiences in very effective ways;

- "suspend the disbelief" of audiences; and

- present pictures of users within the context of everyday experiences.

The use of Actors should not, however, totally replace interactions with "real" users and, in particular, may not be wholly appropriate for very detailed evaluations of a user interface.

The research described above has been focused on older and disabled users, who can present significant challenges to designers, but we believe that these techniques could also be a valuable tool with more traditional user groups, and recommend them for consideration by the HCI community.

Theatre does involves the costs required to script-write, perform, facilitate and record a theatrical session, but these costs are not excessive. They should be compared to the effectiveness, and balanced against, those of other awareness raising and requirement gathering methods. For example, in the study of digital television applications, described above, the costs of using professional theatre were roughly equivalent to six weeks salary of a professional engineer. This was a very small fraction of the total cost of the development project and the data obtained proved very useful, and some of it was unlikely to have been obtained from more traditional requirements gathering exercises.

CHAPTER 13

Attacking the Digital Divide

Engineers, creative designers, advertisers, marketers and policy makers all have a part to play in reducing the Digital Divide. Many older and disabled people, and other minority groups, will remain opposed to using CIT unless systems and interfaces are developed and marketed which are more "friendly" to the digitally disadvantaged. Specialist equipment will remain underused and the benefits of mainstream technological products will not be available to everyone. The focus of this book is the lessons engineers and designers need to learn in order to respond to these challenges.

13.1 THE CHALLENGE

Designers and policy makers often overestimate the challenges and marginal costs of considering the needs of digitally disenfranchised groups. They also can underestimate the benefits, in terms of market share, of the general usability and user friendliness of their designs. Ironically, older people in particular could become a commercial imperative as the conventional markets for CIT products become oversubscribed. Many mainstream CIT designers can be loath to move out of their comfort zone. They, and the companies who employ them, are either fooled by stereotypes and unaware of the market possibilities of these groups, or do not know how to respond to these trends. Some engineers view increasing functionality and novelty as an over-riding consideration, and creative designers often strive for "coolness" at all costs.

Making data on the characteristics of user groups, and guidelines for responding to these, available to designers and engineers is necessary but not sufficient. They also need to develop an attitude of mind that is empathetic to the needs of digitally disadvantaged people. This can require a somewhat different mind-set from that encouraged by traditional CIT markets which tend to focus on young and middle aged, technologically aware, predominantly male users.

This approach can require going against the received truth, both technically and commercially, but such an approach has good antecedents in other fields. History shows that the most exciting innovations occur when a trained and receptive mind is kept open to new opportunities (examples include Penicillin, electricity and x-rays). Designers need the courage to pursue ideas in the face of opposition, but also the wisdom to recognize when they really have got it wrong.

Unfortunately, research funding agencies can be very conservative in their approach. UK research councils can require a statement of what one's research "impact" will be and supporting evidence from potential users and commercial exploiters of the research. This does not encourage really innovative ideas such as those that in the past have produced technologies such as the automatic telephone exchange, the personal computer, and the mobile phone—all of which were initially rejected by most potential users and commercial exploiters. It is unlikely that Bill Gates relied on market research at the early stages of his commercial activities!

13.2   THE CHARACTERISTICS OF OLDER AND DISABLED PEOPLE

There are other inhibiting factors to CIT design for older and disabled people. These users groups are thought to be economically poor and uninterested: in some cases this is correct, but is by no means a universal situation. Their characteristics are perceived to be very different from mainstream users. This again is not universally true. Most people will grow old, their sensory, motor and cognitive abilities declining with this process. Many young and middle-aged people become temporarily disabled due to accident, stress, alcohol or drug (mis)use. Environmental situations, such as low light or high noise, can produce the same effect as a personal disability. The needs of older and disabled people are not different to, they are simply an extension of, those of (what are often called "temporarily") able-bodied people. There are real parallels between the human computer interface needs of disabled (extra-ordinary) people in an ordinary digital environment (e.g., word processing) and ordinary people in an extra-ordinary (e.g., high work load/stress) environment. Designers should exploit these similarities to improve both situations.

Older and disabled people can be considered not to be interested in aesthetics and only concerned with what they "need" to survive, rather than what they "want" to have an interesting life. The effects of these views can be compounded by the users of such equipment not necessarily being the purchasers. This is true for mainstream equipment in the workplace, but unusual for equipment designed for the home. Also, although good subject experts are worth their weight in gold, some teachers, therapists and clinicians may not have the same priorities as older or disabled users, and this should be born in mind when interpreting the views of such professionals.

Requirements depend not only on the disability but also on the personality of the users—some users AAC users always want to sound polite, whereas other users require swear words to be incorporated into their devices. Special equipment for older and disabled people needs to be attuned to the needs and wants of the users. Mainstream technology should not frighten these users and should be as easy to use as is possible.

When considering the design of such equipment we should dream, and dream about "what is possible and desirable, rather than what is feasible (current technology) and what needs to be done (current problems)" [Hollnegal, E., 1991].

> *If you don't have a dream, How are you going to make your dream come true?*
> **Rogers & Hammerstein**

13.3   BUILD ON CAPACITIES

Design for older and disabled users often focuses on the disabilities and how to alleviate them. In contrast, we should look at the abilities of our users and address how we can devise technology that builds on the existing capacities of our users to give them a fuller life.

13.4   BEWARE OF JET PACKS

John Hockenberry, the American journalist and author, has talked about people who offered him "jet packs" as solutions, i.e., complex high-tech devices which were not only unrealistic but did not solve the problems he had. High-tech devices clearly have a place but only if they are an appropriate approach to a challenge. It is important to determine what exactly the challenge is and its priority for users (e.g., when discussing navigation aids, a colleague of mine, Paul Brown, who is blind, has said that he does not need support to find his hotel—the real problem is finding the way back to his room from the hotel dining room).

13.5   TECHNOLOGY-LED RESEARCH

There are situations where technology led research can be very appropriate. This book contains examples of research that was not user led, and some cases where there was real antagonism from clinicians to the concepts being developed. In all cases, however, the researchers did attempt to create and retain an empathy with the users, rather than simply develop the technology because it was fun to develop.

13.6   EXPLOIT UNCONNECTED RESEARCH

Chance favors the prepared mind.

Researchers should be aware of technological advances in other fields rather than just focusing on assistive technology. Even those technologies that have not achieved their promise in the mainstream may have applications within this field. Automatic speech recognition and natural language technology are two examples where a sub-optimal solution in the domain of the research was found to be more than adequate for specific applications for older and disabled people.

"Failures" can lead to success.

13.7 RELATIONSHIPS WITH USERS

If at all possible, designers should develop a long-term relationship with users. It is rarely sufficient to use them only in formal experiments and evaluations. They should be given the opportunity to provide ideas as part of the design process. Having older or disabled people on the research team is ideal but long-term relationships with groups of users, who the researchers can get to know personally, are also very worthwhile. Meeting users in their own environments, in their own homes, or, for example, acting as "helpers" in a Day Centre can provide a wealth of data that would not be apparent from formal meetings in laboratory conditions. Professional theatre can also be valuable for communicating the characteristics of older and disabled people to designers, and facilitating discussions with users and older people. This approach can be particularly beneficial in the conceptual stages of projects. The use of theatre, with its ability to "suspend disbelief", provides opportunities to demonstrate systems in action before they are actually developed.

13.8 HOSTILITY

It is important to encourage vision in researchers, but they also may need protecting. If a researcher is introducing ideas from other disciplines, for example, this may be contrary to received wisdom in both fields and thus disturbing. Unless an idea creates at least some skepticism, or even hostility, it is unlikely to be really novel. It is impossible to predict losers, but winners may not be recognized in their early stages of development. Motorcars were first viewed by most as, at best, dangerous and useless and, at worst, an invention of the devil.

13.9 INTERDISCIPLINARY TEAMS

The vast majority of research in this field requires an interdisciplinary approach and it is necessary to determine the disciplines needed and whether discipline experts should be an integral part of the research team, or only act as advisors to the team. In Dundee we tended to employ subject experts as part of the team: these have included computer scientists and engineers, creative designers, psychologists, therapists, special education teachers, and staff who have benefited from interdisciplinary careers (e.g., psychology and computing, history and computing, and social work and computing). This may not be feasible in smaller research groups and subject experts will have to have a consultant role.

It is important for clinical experts to realize that there are differences between research and clinical support of patients. The primary goal of research is to develop systems for the future and not provide day-to-day support of individual users. This potential conflict can be particularly challenging when technological research is conducted within a clinical environment. Day-to-day support issues are almost always seen as more urgent. It is thus important for practitioners to allocate some dedicated time to research. In contrast, research in an academic environment has the danger of becoming isolated from users and the situations in which they can find themselves.

Do not sacrifice the future for the present.

There are inevitable tensions within an interdisciplinary research group and it is vital that all members realize their own inadequacies and are seen to value the skills of other professionals in the group. This implies that they do not use jargon or "confusion with science" to attack other members of the group. It can be very valuable for the leader, or at least one member of the team, to be prepared to sacrifice time they could spend updating knowledge of their own profession to gain wider view of the field. A computer scientist, for example, could develop a working knowledge of psychology, speech therapy, and sociology. The combination of this, with an understanding of what computer science can and cannot deliver, is a very valuable attribute in a researcher.

13.10 THE NEED FOR PROFESSIONALISM IN THIS RESEARCH AND DEVELOPMENT

It should not be necessary to make the point that the same level of professionalism in research and development is needed for this as for mainstream CIT. There is evidence, however, that some people researching into systems to support older and disabled people do not seem to bring the same level of professionalism to this activity as they do to other areas of work. In the case of academics, this can be manifested by an ignorance of the literature, and some commercial companies seem to believe that equipment for older and disabled people can be of lower standard—in technical, reliability and aesthetic terms—than their mainstream products.

13.11 COGNITIVE DECLINE

Much research has been focused on designing CIT to respond to the motor, and sensory characteristics of older and disabled users, and more needs to be done. There have also been significant research efforts into the needs of people with cognitive disabilities. More attention, however, needs to be paid to the requirements of the older user whose memory and cognitive processes have somewhat declined, and who also have minor sensory impairments. This is a complex situation, but one worthy of in-depth research.

13.12 EVALUATION TECHNIQUES

Clearly, it is important to evaluate the outcomes of our research and development, but the methods adopted in other scientific and clinical research may not be wholly appropriate. This field can see a tension between science and creativity, and between medicine and sociology. Assistive Technology is in a difficult position—the medical authorities see it as medical equipment, whereas most of it is in effect domestic equipment, which would normally have very different evaluation criteria (often only the market share!). It can be difficult, if not impossible, to provide as much evidence for assistive technology systems as is required for a drug trial (e.g., the protocol for a double-blind random controlled study of the efficacy of wheelchairs would be a challenge). I have had a research project for an AAC device turned down because I did not have 100 users in the evaluation (this being more potential users than in the whole of Scotland for that particular piece of equipment!). We should look for, and be prepared to justify, our evaluation techniques within the context of the type of equipment we are developing.

The use of "control" groups can also create problems. One group of researchers were designing PDA based memory aids for people with traumatic brain injury. They wanted to compare them with a control group using pencil and paper reminders. The training time required to use the PDA was much greater than that required for pencil and paper and, in order to have a "proper" control study, they reduced the training time for the PDA based device. The protocol they decided to use thus compared over-trained users of pencil and paper with under-trained users of the PDA system: although this followed strict control study methodology, it was a particularly invalid assessment of the relative efficacy of the two types of system.

> *Observation and judgment can be as valuable as measurement and counting.*
> **(David Crystal Isaac, 1985)**

An overall evaluation of an AAC device has more in common with the evaluation of a speech, poetry or a play than a drug regime. Easily measured criteria, such as keystrokes and words per minutes, tell only a very small part of the story. It is more important to use evaluation methods appropriate to human communication than try to obtain pseudo scientific respectability by using inappropriate evaluation techniques. A very sensitive evaluation measure could, for example, be the level of complaint when a piece of research equipment is removed from the user—although there are ethical challenges in this approach.

13.13 OTHER DIGITALLY DISADVANTAGED GROUPS

The research in this book has been focused on people with particular types of disabilities, certain types of assistive technologies, and the challenges older people have with mainstream technology. The lessons learnt from this research, however, apply to other technologies and users with many other disabilities, and also to other groups who are disenfranchised by culture or educational backgrounds.

13.14 THE CHALLENGES FACED BY OLDER PEOPLE WILL NOT GO AWAY

It is often suggested that the challenges CIT provide to older people, laid out in Chapter 8, are short term. It is argued that, as the current generation of middle-aged people are fully aware of "new" technology, they will have no problems with using it when they become old. This ignores, firstly, the fact that:

- Not all middle-aged people are conversant with "new" technologies.

More importantly, it assumes that:

- people's characteristics will not change as they get older, a very unlikely assumption for the vast majority; and

- the technologies will not change with time.

I fantasize about two old people in the 2020's saying to one another "I could even cope with VISTA, but I am rather scared of these new fangled touch/brain/gesture controlled devices—I cannot understand or remember how to use them".

13.15 CONCLUDING THOUGHTS

It is important for researchers to have the following:

- an in-depth knowledge of the research field;

- a wide perspective, including other research fields which may only be tangentially related; and

- a healthy skepticism of current practice.

They should:

- follow their hunches but keep their feet on the ground;

- be wary of condemning new ideas—it is too easy to dismiss a radical solution and de-motivate a good researcher;

- be fully familiar with, but not dominated by, current service needs;

- spend more time thinking about what they want to achieve than how they are going to achieve it (this is particularly important with the increasing ease of software development);

- have frequent contact with potential users, not only as "experimental subjects", but also as people and colleagues, to improve empathy and the intuitive skills of the researcher.

Design for older and disabled people and most people will be able to use your systems. If you design for young and middle-aged people, you will disenfranchise many minority groups.

He who dares wins.
Special Air Services

Figure 13.1: Design for all ages.

Bibliography

PDFs of a selection of Alan Newell's papers can be found at
www.dundee.computing.ac.uk/staff/afn

Alm, N. (1985) Unpublished MSc Thesis, University of Dundee. Cited on page(s) 47

Alm, N. (1987) Towards a true conversational aid. *Bulletin of the College of Speech Therapists*, March 15–16. Cited on page(s) 47

Alm, N., Arnott, J.L., and Newell, A.F. (1988) Pragmatic issues in the design of a conversation prosthesis. *Proc. North Sea Conf. – Biomedical Engineering*, Maastricht, April. Cited on page(s) 53

Alm, N., Newell, A.F., and Arnott, J.L. (1989) Revolutionary communication system to aid non-speakers. *Speech Therapy In Practice*, **4**(7), vii–viii, March. Cited on page(s) 56

Alm, N., Arnott, J.L., and Newell, A.F. (1992) Evaluation of a text-based communication system for increasing conversational participation and control. In *Proc. of RESNA Intl. '92 Conference "Technology for Consumers,"* Toronto, Canada, 355–368, June. Cited on page(s) 64

Alm, N. (1993) The development of augmentative and alternative communication to assist with social communication. *Technology and Disability*, **2**(3). Cited on page(s) 51

Alm, N., Nicol, M., and Arnott, J.L. (1993) The application of fuzzy set theory to storage and retrieval of conversational texts. Binion, M., Ed., *Proc. RESNA*, The RESNA Press, Washington, DC, 27–129. Cited on page(s) 64

Alm, N. (1994) Ethical issues in AAC research. In Brodin and Bjorck-Akesson, Eds., *Proc. of 3rd ISAAC Research Symposium*, Kerkrade, The Netherlands, Oct. 14–15, Jonkoping University Press. Cited on page(s) 76, 78, 79, 80

Alm, N. and Parnes, P. (1995) Augmentative and alternative communication: Past present and future. *Folia Phoniatr Logop*, **47**, 165–192. DOI: 10.1159/000266349 Cited on page(s) 39, 70

Alm, N. and Newell, A.F. (1996) Being an interesting conversation partner. In *Augmentative and Alternative Communication, European Perspectives*, S. von Tetzchner and M. Jensen, Eds., 171–181. Cited on page(s) 69

Alm, N., Gregor, P., and Newell, A.F. (2002) "Older people and information technology are ideal partners." *International Conference for Universal Design*, Japan, 754–760. Cited on page(s) 83

Alm, N., Dye, R., Gowans, G., Campbell, J., Astell, A., and Ellis, M. (2007) A communication support system for older people with dementia. *IEEE Computer*, **40**(5), 35–41. DOI: 10.1109/MC.2007.153 Cited on page(s) 90

Alm, N., Astell, A., Gowans, G., Dye, R., Ellis, M., Vaughan, P., and Riley, P. (2009) Engaging multimedia leisure for people with dementia. *Gerontechnology*, **8**(4), 236–246. DOI: 10.4017/gt.2009.08.04.006.00 Cited on page(s) 92

Andreasen, P.N., Waller, A., and Gregor, P. (1998) BlissWord—full access to Blissymbols for all users. *Proceedings of the 8th Biennial Conference of ISAAC*, Dublin, Ireland, 167–168. Cited on page(s) 68

Arnott, J.L., Newell, A.F., and Downton, A.C. (1979) A comparison of palantype and stenograph for use in a speech transcription aid for the deaf. *J. Biomedical Engineering*, **1**, 201–210. DOI: 10.1016/0141-5425(79)90042-6 Cited on page(s)

Arnott, J.L., Alm, N., and Newell, A.F. (1988) A text database as a communication prosthesis. *Proc. ICAART 88*, Montreal, Canada, 76–77, June. Cited on page(s) 56, 60

Arnott, J.L. and Javed, M.Y. (1990) Small text corpora in character disambiguation for reduced typing keyboards. *Proc. RESNA*, Washington, DC, 181–182, June. Cited on page(s) 45

Arnott, J.L., Javed, M.Y., and Brophy, B. (1992a) An evaluation of reduced keyboard typing by cerebral palsied adults. *Proc. RESNA*, 296–298, June. Cited on page(s) 45

Arnott, J.L. and Javed, M.Y. (1992b) Probabilistic character disambiguation for reduced keyboards using small text samples. *Augmentative and Alternative Communication*, **8**, 215–223, Sept. DOI: 10.1080/07434619212331276203 Cited on page(s) 45

Ashley, J. (1973) *Journey into Silence*. Bodley Head, London. Cited on page(s) 21

Ashley, J. (1992) *Acts of Defiance*. Penguin, London, 338–339. Cited on page(s) 21, 24, 26

Baker (1966) Final report of the Baker Committee on the Mechanical Recording of Court Proceeding. Hon. Mr Justice Baker, HMSO, Sept. Cited on page(s) 27

Baker, B. (1982) Minspeak. *Byte*, 186–202, Sept. Cited on page(s) 39, 42

Baker, R.G. and Newell, A.F. (1980) Teletext subtitles for the deaf–problems in linguistics and psychology. *Proc. Int. Broadcasting Convention*, London, 97–100. Cited on page(s) 33

Baker, R.G. (1981) *Guidelines for Subtitling Television Programmes*. Independent Broadcasting Authority. Cited on page(s) 33

Baker, R.G., Downton, A.C., and Newell, A.F. (1981) Simultaneous speech transcription and television captions for the deaf. *Proc. of Visible Language,* **2**, P.A. Kolers, M.E. Wrolstad, and H. Bouma, Eds., Plenum Pub. Corp. Cited on page(s) 33

Beattie, G. (1979) Planning units in spontaneous speech. *Linguistics,* **17**, 61–78. DOI: 10.1515/ling.1979.17.1-2.61 Cited on page(s) 53

Becker, H.A. (2000) Discontinuous change and generational contracts. In S. Arber and C. Attiats-Donfut, Eds., *The Myth of Generational Conflict,* Routledge and Kegan-Paul, UK, 114–132. Cited on page(s) 103

Beukelman, D.R. and Mirenda, P. (1992) *Augmentative and Alternative Communication: Management of Severe Communication Disorders in Children and Adults.* Paul H. Brookes, London. Cited on page(s) 37

Beukelman, D.R. and Mirenda, P. (1998) *Augmentative and Alternative Communication Management of Severe Communication Disorders in Children and Adults.* 2nd ed., Brookes, Baltimore, MD. Cited on page(s) 67

Benyon, D. and Macaulay, C. (2002) Scenarios and the HCI-SE design problem. *Interacting with Computers,* **14**, 397–405. DOI: 10.1016/S0953-5438(02)00007-3 Cited on page(s) 129

Birdwhistle, R. (1974) The language of the body. In Sivverstone, A., Ed., *Human Communication,* Wiley, New York, 203–220. Cited on page(s) 52

Black, A., Waller, A., Pullin, G., Abel, E. (2008) "Introducing the phonic-stick: Preliminary evaluation with seven children." *ISAAC 2008, 13th Biennial Conference of the International Society for Augmentative and Alternative Communication,* Montreal, Canada, 2–7 August, 424. Cited on page(s) 70

Black, R., Reddington, J., Reiter, E., Tintarev, N., and Waller, A. (2010) Using NLG and sensors to support personal narrative for children with complex communication needs. First Workshop on Speech and Language Processing for Assistive Technologies (SLPAT), *Human Language Technologies: The 11th Annual Conference of the North American Chapter of the Association for Computational Linguistics,* Los Angeles. Cited on page(s) 67

Bliss, C. (1965) *Semantography (Blissymbolics).* Semantography Publications, Sidney. Cited on page(s) 67

Blomberg, J., Burrell, J.M., and Guest, G. (2002) An ethnographic approach to design. In Jacko, J. and Sears, A., Eds., *The Human-Computer Interaction Handbook,* Lawrence Erlbaum, Mahwah, New Jersey, 964–986. Cited on page(s) 118

Blythe, G. (2002) Optimising uptake of usability recommendations. In *Proceedings of the 1st European Usability Professionals Association Conference,* London, UK. Cited on page(s) 115

156 BIBLIOGRAPHY

Boal, A. (1995) *The Rainbow of Desire*. Routledge, London. Cited on page(s) 130

Booth, A.W. and Barnden, M.S. (1979) Voice input to English text output. *Int. J. Man–Machine Studies*, **11**, 681–691. DOI: 10.1016/S0020-7373(79)80024-3 Cited on page(s) 33

Brooks, C.P. and Newell, A.F. (1985) Computer transcription of handwritten shorthand as an aid for the deaf—a feasibility study. *Int. J. of Man–Machine Studies*, **23**(1), 45–60. DOI: 10.1016/S0020-7373(85)80023-7 Cited on page(s) 27

Brophy-Arnott, M.B., Newell, A.F., Arnott, J.L., and Condie, D. (1992) A survey of the communication impaired population of Tayside. *European Journal of Disorders of Communication*, **27**(2), 159–173. DOI: 10.3109/13682829209012037 Cited on page(s) 47

Broumley, L., Arnott, J.L., Cairns, A.Y., and Newell, A.F. (1990) TalksBack: An application of AI techniques to a communication prosthesis for the non-speaking. *Proc. European AI Conf.*, Stockholm, 117–119, August. Cited on page(s) 60

Bruno, J. (1989) Customizing a Minspeak system for a preliterate child. *Augmentative and Alternative Communication*, **5**, 89–100. DOI: 10.1080/07434618912331275066 Cited on page(s) 67

Cahn, J.E. (1988) "From sad to glad," emotional computer voices. *Proc. Speech Tech. '88*, New York, 35–36. Cited on page(s) 70

Cahn, J.E. (1989) Generation of affect in synthesized speech. *Proc. American Voice Input/Output Society*. Cited on page(s) 70

Cairns, A.Y., Peddie, H., Cobley, A., Arnott, J.L., and Newell, A.F. (1993) Multi-modal computer interaction for the physically disabled. In *Proc. of ECART 2*, 2nd European Conf. on the Advancement of Rehabilitation Technology, Stockholm, 62–64, May. Cited on page(s) 112

Carmichael, A. (1999) *Style guide for the design of interactive television services for elderly viewers*. Independent Television Commission, Kings Worthy Court, Winchester, UK. Cited on page(s) 101

Carmichael, A., Newell, A.F., and Morgan, M. (2007) "The efficacy of narrative video for raising awareness in ICT designers about older users' requirements." *Interacting with Computers*, **19**, 587–596. DOI: 10.1016/j.intcom.2007.06.001 Cited on page(s) 134

Carroll, J.M. and Carrithers, C. (1984) Training wheels in a user interface. *Communications of the ACM*, **27**(8), 800–806. DOI: 10.1145/358198.358218 Cited on page(s) 87

Cassim, J. (2005) Challenge. (Annual publication), Helen Hamlyn Centre, London. Cited on page(s) 123

Cherns, A. (1976) Principles of socio-technical design. *Human Relations*, **29**(8), 783–792. DOI: 10.1177/001872677602900806 Cited on page(s) 89

Clark, C.R. (1984) A close look at the standard Rebus system and Blisssymbolics. *J. of Ass. for Persons with severe Handicapps,* **9**, 37–48. Cited on page(s) 67

Clarke, H.H. and Clarke, E.V. (1977) *Psychology and Language.* Harcourt Brace Jovanovich, New York. Cited on page(s) 48

Cohen, G. (2001) *The Creative Age: Awakening Human Potential in the Second Half of Life.* Avon Book, New York. Cited on page(s) 98

Cohen, G. (2005) *The Mature Mind.* Basic Books, New York. Cited on page(s) 102

Colby, K.M., Christinaz, D.U., Parkinson, R., and Tiedman, M. (1982) Predicting word-expressions to increase output rates with speech prostheses used in communication disorders. *Proc. IEEE Int. Conf. on acoustics speech and signal processing,* **22**, 751–754, May. Cited on page(s) 40

Colwell, C. and Petrie, H. (1999) Evaluation of guidelines for designing accessible web content. In C. Buhler and H. Knops, Eds., *Assistive Technology on the Threshold of the New Millennium, (Proceedings of AAATE '99).* IOS Press, Amsterdam. Cited on page(s) 118

Cooper, A. (1999) *The Inmates are Running the Asylum.* Macmillan, USA. Cited on page(s) 129

Copeland, K. (1974) *Aids for the Severely Handicapped.* Sector Publishing, London. Cited on page(s) 16, 37

Copestake, A., Langer, S., Paluzuelos Cagigas, S., Eds. (1997) Natural language processing for communication aids. *Proc. of workshop at 35th Annual Meeting of the Association of Computational Linguistics,* Madrid Spain. Cited on page(s) 43

Cushler, C., Badman, A., Demasco, P., and McCoy, K. (1996) A communication aid enhanced with semantic parsing. *Proc. 7th Ann. Conf. of Int. Soc. of Augmentative and Alternative Communication,* Vancouver, Canada, 493–494. Cited on page(s) 39

Czaja, S.J. and Lee, C.C. (2008) "Information technology and older adults." In *The Human Computer Interaction Handbook,* Sears, A. and Jacko, J., Eds., Lawrence Erlbaum, New Jersey, 777–792. Cited on page(s) 98, 99, 100, 101

Danowski, J.A. and Sacks, W. (1980) Computer communication and the elderly. *Experimental Aging Research,* **6**(2), 125–135. DOI: 10.1080/03610738008258350 Cited on page(s) 97

Deary, I., Whalley, L.J., and Starr, J.M. (2009) A Lifetime of Intelligence: Follow-up Studies of the Scottish Mental Surveys of 1932 and 1947. American Psychological Association. Cited on page(s) 102

Demasco, P.W. and McCoy, K.F. (1992) Generating text from compressed input: An intelligent interface for people with severe motor impairment. *Communications of the ACM,* **35**(5), 68–79. DOI: 10.1145/129875.129881 Cited on page(s) 39

Dickinson, A., Eisma, R., Syme, A., and Gregor, P. (2002) Utopia: usable technology for older people: inclusive and appropriate. In *A New Research Agenda for Older Adults, Proceedings of the 16th British HCI Conference*, London, S. Brewster and M. Zajicek, Eds., **2**, 38–39, Sept. Cited on page(s) 99, 120

Dickinson, A., Goodman, J., Syme, A., Eisma, R., Tiwari, L., Mival, O., and Newell, A.F. (2003) Domesticating technology in-home requirements gathering with frail older people. *10th International Conference on Human-Computer Interaction HCI*, 22–27 June, Crete, Greece, C. Stephanidis, Ed., **4**, 827–831. Cited on page(s) 99, 101, 121

Dickinson, A., Newell, A.F., Smith, M.J., and Hill, R. (2005) "Introducing the internet to the over-60's: Developing an email system for older novice computer users." *Interacting with Computers*, **17**, 621–642. DOI: 10.1016/j.intcom.2005.09.003 Cited on page(s) 85

Dishman, E. (2003) Designing for the New World. In *Design Research*, Laurel, B., Ed., MIT Press, 41–48. Cited on page(s) 130

Dunne, A. (1999) *Hertzian Tales: Electronic products, aesthetic experience and critical design.* RCA CRD Research, new edition published by the MIT Press, 2006. Cited on page(s) 126

Dunne, A. and Raby, F. (2001) *Design Noir–The Secret Life of Electronic Objects.* Birkhäuser. Cited on page(s) 127

Dye, R., Alm, N., Arnott, J.L., Harper, G., and Morrison, A.I. (1998) A script-based AAC system for transactional interaction. *Natural Language Engineering*, **4**, 57–71. DOI: 10.1017/S1351324998001867 Cited on page(s) 64

Edmundson, W. (1981) *Spoken Discourse, a Model for Analysis.* Longman, London, UK. Cited on page(s) 53

Eisma, R., Dickinson, A., Goodman, J., Mival, O., Syme, A., and Tiwari, L. (2002) Mutual inspiration in the development of new technology for older people. *New Research Agenda for Older Adults, BCS HCI 2002*, London, S. Brewster and M. Zajicek, Eds., 38–43. Cited on page(s) 120

Eisma, R., Dickinson, A., Goodman, J., Mival, O., Syme, A., and Tiwari, L. (2003) Mutual inspiration in the development of new technology for older people. In *Proc. Include 2003*, March, London, 7:252–7:259. Cited on page(s) 99, 123

Eisma, R., Dickinson, A., Goodman, J., Syme, A., Tiwari, L., and Newell, A.F. (2004) Early user involvement in the development of information technology related products for older people. *International Journal Universal Access in the Information Society*, **3**(2). DOI: 10.1007/s10209-004-0092-z Cited on page(s) 121, 123

Ellis, A.W. and Young, A.W. (1988) *Human Cognitive Neuropsychology.* Earlbaum Associates, London. Cited on page(s) 42

Etchels, M.C., MacAulay, F., Judson, A., Ashraf, S., Ricketts, I.W., Waller, A., Alm, N., Warden, A., Gordon, B., Brodie, J., and Shearer, A.J. (2003) ICU-Talk: the development of a computerised communication aid for patients in intensive care. *Care of the Critically Ill,* **19**(1), 4–9. Cited on page(s) 68

Fisk, A.D., Rogers, W., Charness, N., Czaja, S.J., and Sharit, J. (2004) *Designing for Older Adults: Principles and Creative Human Factors Approaches.* Taylor and Francis, London. Cited on page(s) 100

Foulds, R.A., Baletsa, G., and Crochetiere, W.J. (1975) The effectiveness of language redundancy in non-verbal communication. *Proc. Conf. on Devices and Systems for the Disabled,* Philadelphia, 82–86. Cited on page(s) 16

Foulds, R., Soede, M., van Balkom, H., and Boves, L. (1987) Lexical prediction techniques applied to reduce motor requirements for augmentative communication. *Proc. RESNA,* San Jose, California, USA, 115–117. Cited on page(s) 45

Freudenthal, D. (1997) Learning to use interactive devices, age differences in reasoning processes. Unpublished Masters Thesis, Eindhoven University of Technology, Netherlands. Cited on page(s) 100

Frick, R.W. (1985) *The Communication of Emotional Meaning.* McGraw Hill, New York. Cited on page(s) 69

Gajos, K.Z., Wobbrock, J.O., and Weld, D.S. (2008) Improving the performance of motor-impaired users with automatically-generated, ability-based interfaces. *Proceedings of the ACM Conference on Human Factors in Computing Systems (CHI '08),* Florence, Italy (April 5–10), ACM Press, New York, 1257–1266. DOI: 10.1145/1357054.1357250 Cited on page(s) 75

Garfinkel, H. (1967) *Studies in Ethomethodology.* Prentice Hall, Englewood Cliffs, NJ. Cited on page(s) 60

Gheeraw, R.R. and Lebbon, C.S. (2002) Inclusive design – Developing theory through practice. In *Universal Access and Assistive Technology,* Keates, S., Langdon, P.M., Clarkson, P.J., and Robinson, P., Eds., Springer-Verlag, London, 43–52. Cited on page(s) 121

Gilligan, R., Campbell, P., Dries, J., and Obermaier, N. (1998) The current barriers for older people in accessing the information society. European Institute for the Media, Düsseldorf and the Netherlands Platform Older People and Europe, Utrecht. Cited on page(s) 97

Goffman, E. (1981) *Forms of Talk.* Basil Blackwell, Oxford, UK. Cited on page(s) 53

Goodman, J., Syme, A., and Eisma, R. (2003) Age-old Question(naire)s. *Include 2003,* Helen Hamlyn Institute, London. Cited on page(s) 98

Grandstrom, B. (1990) Speech technology cross fertilization between research for the disabled and the non-disabled. *Proc. Int. Soc. Augmentative and Alternative Communication, Research Symposium,* Swedish Handicap Institute, Stockholm, Sweden, 131–134. Cited on page(s) 107

Gregor, P., Alm, N., Arnott, J.L., and Newell, A.F. (1999) "The application of computing technology to interpersonal communication at the University of Dundee's Department of Applied Computing." *Technology and Disability,* 107–113. Cited on page(s) 6

Gregor, P. and Newell, A.F. (2001) "Designing for dynamic diversity–making accessible interfaces for older people." *WUAUC'01, ACM,* J. Jorge, Ed. DOI: 10.1145/564526.564550 Cited on page(s)

Gregor, P., Newell, A.F., and Zajicek, M. (2002) "Designing for dynamic diversity–interfaces for older people." *ASSETS 2002,* The Fifth International ACM Conference on Assistive Technologies, 8–10 July, Edinburgh, Scotland. J.A. Jacko, Ed., 151–156, ISBN: 1 58113 464 9. DOI: 10.1145/638249.638277 Cited on page(s) 105

Gregor, P., Dickinson, A., Mcaffer, A., and Andreasen, P. (2003) See-Word – a personal word processing environment for dyslexic computer users. *British Journal of Educational Technology,* **34**(3), 341–355. DOI: 10.1111/1467-8535.00331 Cited on page(s) 44

Gregor, P., Sloan, D., and Newell, A.F. (2005) "Disability and technology: Building barriers or creating opportunities?" *Advances in Computers,* **64**, 283–346. DOI: 10.1016/S0065-2458(04)64007-1 Cited on page(s)

Grudin, J. (2006) Why personas work–the psychological evidence. In *The Persona Lifecycle, Keeping People in Mind throughout Product Design.* Pruitt, J. and Adlin, T., Eds., Elsevier, 642–663. Cited on page(s) 129

Guenthner, F., Langer, S., Kruger-Thielmann, K., Pasero, R., and Sabatier, P. (1993) KOMBE, Communication aids for the handicapped Munxhen. *CIS Report,* 92–55. Cited on page(s) 43

Gumperz, J. (1982) *Discourse Strategies,* Cambridge University Press, UK, 133. Cited on page(s) 48, 53, 59

Hales, G. (1976) Communicating with the deaf by conventional orthography. *Brit. J. Audiol.,* **10**(3), 83–86. DOI: 10.3109/03005367609078813 Cited on page(s) 21

Hawthorn, D. (2000) Possible implications of aging for interface designers. *Interacting with Computers,* **12**, 507–528. DOI: 10.1016/S0953-5438(99)00021-1 Cited on page(s) 88, 103

Hawthorn, D. (2002) Designing usable applications for older users–an example. *Human Factors,* Melbourne. Cited on page(s) 84, 87, 99, 101

Hawthorn, D. (2003) How universal is good design for older users? *Proc. ACM Conf. on Universal Usability,* Vancouver, Canada, 38–47, Nov. DOI: 10.1145/957205.957213 Cited on page(s) 88, 103

Hawthorn, D. (2006) Designing Effective Interfaces for older users. PhD Thesis, Univ. of Waikato, New Zealand. Cited on page(s) 88

Hayward, G. (1979) A profoundly deaf businessman's views on the Palantype Speech Transcription System. *International Journal of Man-Machine Studies,* **11**, 711–715, Nov. DOI: 10.1016/S0020-7373(77)80023-0 Cited on page(s) 27

Head, A.J. (2003) Personas: Setting the stage for building usable information sites. *Information Today/Online,* **27**(4), 14–21, July-Aug. Cited on page(s) 129

Heckathorne, C.W. and Childress, D.S. (1983) Applying Anticipatory Text selection in a writing aid for people with severe motor impairment. *IEEE Micro,* 17–23, June. DOI: 10.1109/MM.1983.291116 Cited on page(s) 39

Helander, M., Landauer, T.K., and Prabhu, P., Eds. (1997) *Handbook of Human-Computer Interaction.* Elsevier Science BV, 813–824. Cited on page(s) 118

Heller, R., Jorge, J., and Guedj, R., Eds. (2001) *EC/NSF Workshop on Universal Accessibility of Ubiquitous Computing: Providing for the Elderly,* 22–25 May, Portugal, ISBN: 1 58113 424X, 90–92. DOI: 10.1145/564526.564528 Cited on page(s)

Hickey, M. and Page, C.J. (1993) Polyvox: flexible message selection in a communication prosthesis for non-speakers. *Proc. 2nd Euro Conf. on the advancement of Rehabilitation Technology,* Stockholm, Sweden, 26–28, May. Cited on page(s) 64

Higgingbotham, D.J. (1990) Evaluation of keystroke savings across five assistive communication technologies. *Augmentative and Alternative Communication,* **8**, 258–272. DOI: 10.1080/07434619212331276303 Cited on page(s) 40

Higginbotham, D.J., Shane, H., Russel, S., and Caves, K. (2007) Access to AAC: Present, Past, and Future. *Augmentative and Alternative Communication,* **23**(3), 243–257, Sep. DOI: 10.1080/07434610701571058 Cited on page(s) 65, 70

HMSO (1977) Third Report of Select Committee of the House of Commons (Services), Session 1976–77, HMSO. Cited on page(s) 22

Hocking, C. (1999) Function or feelings: factors in abandonment of assistive devices. *Technology and Disability,* **11**, 3–11. Cited on page(s) 125

Hoey, J., Poupart, P., von Bertoldi, A., Craig, T., Boutilier, C., and Mihailidis, A. (2010) Automated handwashing assistance for persons with dementia using video and a partially observable Markov decision process. *Computer Vision and Image Understanding,* **114**(5). DOI: 10.1016/j.cviu.2009.06.008 Cited on page(s) 92, 142

Hollnegal, E. (1991) MINES, ParisTech, Private Communication. Cited on page(s) 146

Howard, J.H., Jr. and Howard, D.V. (1997) Learning and memory. In A.D. Fisk and W.A. Rogers, Eds., *Handbook of Human Factors and the Older Adult*, San Diego, CA, Academic Press, 7–26. Cited on page(s) 103

Hunnicutt, S. (1989) Using syntactic and semantic information in a word prediction aid. *Proc. Eurospeech*, Paris, France, 191–193. Cited on page(s) 43

Hypponen, H. (1999) *The Handbook on Inclusive Design for Telematics Applications*. Siltasaarenkatu 18A, 00531 Helsinki, Finland. Cited on page(s) 116

Inglis, E., Szymkowiak, A., Gregor, P., Newell, A.F., Hine, N., Wilson, N.A., and Evans, J. (2002) Issues surrounding the user-centred development of a new interactive memory aid. In *Universal Access and Assistive Technology, (Proceedings of the Cambridge Workshop on UA and AT ' 02 2002)*, S. Keates, P. Langdon, P. John Clarkson, and P. Robinson, Eds., 171–178. Cited on page(s) 89

ISO 13407 (1999) Human-centered design processes for interactive systems. International Organisation for Standardization. Cited on page(s) 116

Jefferson, G. (1984) On stepwise transition from talk about trouble to inappropriately next-positioned matters. In Atkins, J. and Heritage, J., Eds., *Structures of Social Action–Studies in Conversation Analysis*, Cambridge University Press, London, 191–222. Cited on page(s) 64

Jessome, J. and Parks, C. (2001) Everyday Technology and Older Adults: Friends or Foes? Nova Scotia Centre on Aging, Mount Saint Vincent University. Cited on page(s) 97

Johnson, J.M., Inglebret, E., Jones, C., and Ray, J. (2006) Perspectives of speech language pathologists regarding success versus abandonment of AAC. *Augmentative and Alternative Communication*, **22**(2), 85–99. DOI: 10.1080/07434610500483588 Cited on page(s)

Johnson, R. (1981) *The Picture Communication Symbols*, book 1. Mayer-Johnston, Solana Beach, CA, USA. Cited on page(s) 67

Judge, S. and Towend, G. (2010) User Perceptions of Communication Aid Design. D4D AAC Project Report, Devices for Dignity, Barnsley Hospital and Sheffield PCT, UK. Cited on page(s) 71

Kamphuis, H.A., and Soede, M. (1989) KATDAS, A small number of keys direct access system. *Proc. RESNA*, New Orleans, USA, 278–279, June. Cited on page(s) 45

Keates, S. and Clarkson, P.J. (2003) "Countering design exclusion: Bridging the gap between usability and accessibility." *Universal Access in the Information Society*, **2**(3), 215–225. DOI: 10.1007/s10209-003-0059-5 Cited on page(s) 116

Keates, S. and Clarkson, J. (2004) *Countering Design Exclusion–An Introduction to Inclusive Design*. Springer. Cited on page(s) 116

Kline, D.W. and Schieber, F.J. (1995) Vision and ageing. In Birren, J.E. and Schaie, K.W., Eds., *Handbook of the Psychology and Ageing*, Van Nostrand Reinhold, New York, 296–331. Cited on page(s) 101

Knight, J. and Jefsioutine, M. (2002) Relating usability to design practice. In *Proceedings of 1st European Usability Professionals Association Conference*, London. Cited on page(s) 115

Korba, L., Nelson, P.J., and Park, G.C. (1985) Morse code computer input using the MOD keyboard. *Proc. RESNA 8th Conference*, Memphis, TN, USA, 58–60. Cited on page(s) 37

Kosinski, J. (1968) *Being there*. New York, Harcourt Brace, Jovanovich. Cited on page(s) 49

Koskinen, I., Battarbee, K., and Mattelmaki, T., Eds. (2003) *Empathetic Design: User Experience in Product Design*, IT Press, Helsinki, Finland. Cited on page(s) 118

Kraat, A. (1985) State of the art report–communication between aided and natural speakers (IPCAS report), Toronto. Canadian Rehabilitation Council for the disabled. Cited on page(s) 38, 39

Kraat, A. (1990) AAC focus for the '90s: New technologies or consumer use and outcomes? In Beth Mineo, Ed., *Augmentative and Alternative Communication in the next Decade*, University of Delaware, 20–22. Cited on page(s) 71

Kraat, A.W. (1991) Methodological issues in the study of language development amongst children using aided language. In Brodin, J. and Bjork-Akesson, E., Eds., *Methodological Issues in Research in Augmentative and Alternative Communication*, Swedish Handicap Institute, Stockholm. 118–123. Cited on page(s) 67

Lambourne, A.W., and Brooks, C.P., and Baker, R. (1982) Information Transformation in the Visual Presentation of Speech. *Proc. Int. Conf. on Man-Machine Systems*, Manchester, 181–185. Cited on page(s) 5, 34

Lambourne, A.D., King, R.W., and Newell, A.F. (1982) A man-machine system for on-line or off-line preparation of teletext subtitles. *Proc. Intl. Conf. On Man-Machine Systems*, Manchester, 181–185, July. Cited on page(s)

Laver, J. (1974) Communicative functions of phatic communion. In Laver, J., Ed., *Semiotic Aspects of Spoken Conversation*, Edward Arnold, London. Cited on page(s) 48

Lazar, J. (2007) *Universal Usability: Designing Computer Interfaces for Diverse User Populations*. John Wiley and Sons. Cited on page(s) 115

Lehtinen, V., Nasanenm, J., and Sarvas, R. (2010) "A little silly and empty headed." *Interfaces 85*, Winter, 05–07. Cited on page(s) 94

Levine, S.H., Goodenough-Trepagnier, C., Getscho, C.O., Minneman, S.L. (1987) *Proc. RESNA*, San Jose, California, USA, 115–117. Cited on page(s) 45

Levinson, S. (1983) *Pragmatics*. Cambridge University Press, UK. Cited on page(s) 53

Light, J., Collier, B., and Parnes, P. (1985) Communicative interaction between young non-speaking physically disabled children and the primary care givers. *Augmentative and Alternative Communication*, Part I: 1,2, 74–83; Part II: 1,3, 98–107. DOI: 10.1080/07434618512331273591 Cited on page(s) 47

Light, J. (1988) Interaction involving individuals using augmentative and alternative communication systems, state of the art and future directions. *Augmentative and Alternative Communication*, **4**(2). DOI: 10.1080/07434618812331274657 Cited on page(s) 38, 39

Light, J., Parnes, P., Lindsay, P., and Siegel, L. (1990) The effects of message encoding techniques on recall by literate adults using AAC systems. *Augmentative and Alternative Communication*, **6**, 184–202. DOI: 10.1080/07434619012331275454 Cited on page(s) 40, 52

Lim, C.S.C. (2010) Designing inclusive ICT products for older users: taking into account the technology generation effect. *Journal of Engineering Design*, **21**(2–3), 189–206, April–June. DOI: 10.1080/09544820903317001 Cited on page(s) 103

Lowe, H., et al. (1974) Lightwriter. In Copeland, K., Ed., *Aids for the Severely Handicapped*, Sector Press. Cited on page(s) 19

Maharaj, S. (1980) Pictogram ideogram communication. *The Pictogram Centre Saskatchewan Ass. of Rehab. Centres*, Saskatoon SK. Cited on page(s) 67

Mailing, R.G., and Clarkson, D.C. (1963) Electronic Controls for the Tetraplegic. *Paraplegia*, **1**(3). Cited on page(s) 16

Mailing, R.G. (1968) Control in severe disability. *Rehabilitation*, **64**, Jan. Cited on page(s) 16

Maltz, D.N. and Borker, R.A. (1982) A cultural approach to male-female miscommunication. Gumperz, J. Ed., *Language and Social Identity*, Cambridge University Press, UK, 196–216. Cited on page(s) 53

McCoy, E. and Shumway, R. (1979) Real time captioning–promise for the future. *American Annuals of the Deaf*, **125**(5), 681–690. Cited on page(s) 33

McCoy, K. and Demasco, P. (1985) Some applications of natural language processing to the field of augmentative and alternative communication. *Proc. IJCAI '95 workshop on developing AI applications for disabled people*. Montreal, Canada, 97–112. Cited on page(s) 43

McDonald, E.T. (1980) Teaching and using Bliss Symbols. *Blissymbolics Communications Institute*, Torronto, Canada. Cited on page(s) 67

McGaffey, A., Haaf, R., and Millen, N. (1991) Clinicians and a consumer's perspective. *Communicating Together*, **9**(3), 12–13. Cited on page(s) 39, 52

McGregor, A., Filz, G., and Alm, N. (1991) A topic discussion and lecturing aid for non-speakers using hypertext and a graphical user interface. *Proc. 3rd National ISAAC Conference,* Mansfield, England, UK. Cited on page(s) 63

McGregor, A. and Alm, N. (1992) Thoughts of a nonspeaking member of an AAC research team. 5th Int. Conf. ISAAC. Abstract in *Augmentative and Alternative Communication,* **8**(2), Philadelphia, USA. Cited on page(s) 63

McKenna, S.J. and Nait Charif, H. (2005) "Summarising contextual activity and detecting unusual inactivity in a supportive home environment." *Pattern Analysis and Applications,* **7**, 386–401. DOI: 10.1007/s10044-004-0233-2 Cited on page(s) 89

McKinlay, A. (1991) Using a social approach in the development of a communication aid to achieve perceived communication competence. *Proc. 14th RESNA,* Kansas City, MO, USA, 204–206. Cited on page(s) 49, 60

McKinlay, A. and Newell, A.F. (1992) Conversation analysis in AAC. In *Proc. of 2nd ISAAC Research Symposium, "Methodological Issues in Research in Augmentative and Alternative Communication,"* J. Gardner-Bonneau, Ed., Philadelphia, 128–132, August. Cited on page(s) 60

McKinlay, A., Beattie, W., Arnott, J.L. (1995) Augmentative and Alternative Communication. The role of Broadband Telecommunications. *IEEE Trans. on Rehab. Engineering,* **3**(3), 254–260, September. DOI: 10.1109/86.413198 Cited on page(s) 46

McNaughton, S. and Kates, B. (1974) Visual Symbols: communication system for the pre-reading physically handicapped child. *American Ass. On Mental Deficiency Ann. Meeting,* Toronto, Canada. Cited on page(s) 67

McNaughton, K.B. (1980) "The application of Blisssymolics." In Schiefelbusch, R.L., Ed., *Non-speech Language Communication Analysis and Intervention,* University Park Press, Baltimore, MD, 303–321. Cited on page(s) 43

Meiselwitz, G., Wentz, B., and Lazar, J. (2009) Universal Usability: Past, Present, and Future Foundations and Trends. In *Human-Computer Interaction,* **3**(4). DOI: 10.1561/1100000029 Cited on page(s) 116

Mihailidis, A., Boger, J., Candido, M., and Hoey, J. (2008) The COACH prompting system to assist older adults with dementia through handwashing: An efficacy study. *BMC Geriatrics,* **8**(28). DOI: 10.1186/1471-2318-8-28 Cited on page(s) 92

Miller, G. (1956) The magical number seven plus or minus two: some limits on our capacity for processing information. *Psychological Review,* **63**, 81–97. DOI: 10.1037/h0043158 Cited on page(s) 40

Milne S., Dickinson, A., Carmichael, A., Sloan, D., Eisma, E., and Gregor, P. (2005) "Are guidelines enough? An introduction to designing web sites accessible to older people." *IBM Systems Journal,* **44**(3), 557–571. DOI: 10.1147/sj.443.0557 Cited on page(s) 118

Mineo, B., Ed. (1990) *Augmentative and Alternative Communication in the Next Decade.* University of Delaware, A.S.E.L. Cited on page(s) 39

Minniman, S.L. (1985) A simplified touch-tone telecommunications aid for deaf and hearing impaired individuals. *Proc. RESNA,* **85,** 209–211. Cited on page(s) 45

Morgan, M., McGee-Lennon, M.R., Hine, N., Arnott, J.L., Martin, C., Clark, J., and Wolters, M. (2008) Living in a Smart House: from "Requirements gathering with diverse user groups and stakeholders." *ACM SIGCHI Conf. on Human Factors in Computing Systems (CHI),* Extended Abstracts, Florence, Italy, 7th–10th April, 2597–2600. DOI: 10.1145/1358628.1358720 Cited on page(s) 89, 138, 141, 142

Morris, D. (1982) *Manwatching.* Triad/Grenada. Cited on page(s) 38

Morris, C., Newell, A.F., Booth, L., and Arnott, J.L. (1991) Syntax PAL–a system to improve the syntax of those with language dysfunction. In *Proc. of 14th Annual RESNA Conference "Technology for the Nineties,"* Kansas City, USA, 105–106, June. DOI: 10.1080/10400435.1992.10132194 Cited on page(s) 43

Morris, J.M. (1992) The effects of an introductory computer course on the attitudes of older adults towards computers. *Proceedings of the twenty-third SIGCSE technical symposium on Computer science education,* Kansas City, Missouri, USA, 74. DOI: 10.1145/134510.134526 Cited on page(s) 85

Morton, K. (1992) PALM, Psychoacoustic language modeling. *Proc. Inst. Of Acoustics,* **14**(6) 198–197. Cited on page(s) 56

Muller, M.J. (2002) Participatory design: the third space. In *The Human-Computer Interaction Handbook,* Jacko, J.A. and Sears, A., Eds., Lawrence Erlbaum, New Jersey, 1051–1068. Cited on page(s) 118

Murray, I.R., Arnott, J.L., and Newell, A.F. (1988) HAMLET–simulating emotion in synthetic speech. *Proc. Speech '88, 7th FASE Symposium,* Edinburgh, 1217–1223. Cited on page(s) 69

Murray, I.R. and Arnott, J.L. (1993) Toward the simulation of emotion in synthetic speech: A review of the literature on human vocal emotion. *J. Acoustic. Soc. Amer.,* **93**(2), 1097–1108, Febr. DOI: 10.1121/1.405558 Cited on page(s) 69

Myerson, J. (2004) *IDEO: Masters of Innovation.* Laurence King. Cited on page(s) 126

Nair, S.N., Lee, C.C., and Czaja, S.J. (2005) Older Adults and attitudes towards computers. *Proc. 49th Ann. Meet. of the Human Factors and Ergonomics Society,* Orlando, FL, USA, 154–157. Cited on page(s) 100

National Shorthand Reporter (1974) "Transcription by Computer." *National Shorthand Reporter*, **35**(7), April. Cited on page(s) 22

Negoita, C.V. (1976) On fuzzyness in information retrieval. *Int. J. Man-Machine Studies*, **6**(8), 711–716. DOI: 10.1016/S0020-7373(76)80032-6 Cited on page(s) 64

Newell, A.F. and Nabavi, D. (1969) Votem, The voice operated typewriter employing morse code. *J. of Sc. Inst.*, **2**(2), 655–657. DOI: 10.1088/0022-3735/2/8/314 Cited on page(s) 15

Newell, A.F. (1970) Man machine communication using spoken morse code. *Intl. J. of Man-Machine Studies*, **2**, 351–362. DOI: 10.1016/S0020-7373(70)80003-7 Cited on page(s) 15

Newell, A.F. (1974a) *The Talking Brooch in Aids for the Severely Handicapped.* Copeland K., Ed., Sector Pubs. Cited on page(s) 4, 20

Newell, A.F. (1974b) Can speech recognition machines help the deaf? *The Teacher of the Deaf*, **72**, 367–374. Cited on page(s) 22

Newell, A.F., Beynon, J.D.E., Brumfitt, P.J., Houssain, K.S. (1975) An alphanumeric display as a communication aid for the dumb. *Medical and Biological Eng.*, **13**(2), 84–88. DOI: 10.1007/BF02478192 Cited on page(s) 17

Newell, A.F. (1977) Communication aids for people with impaired speech and hearing. *Electronics and Power*, **23**(10), 821–827. DOI: 10.1049/ep.1977.0453 Cited on page(s)

Newell, A.F. and King, J.A.F. (1977) Speech translation systems for the hearing impaired. *Medical and Biological Eng. and Computing*, **15**, 558–563. DOI: 10.1007/BF02442285 Cited on page(s) 24

Newell, A.F. (1979) The relevance of teletext and viewdata to communication disorders. *Child Health, Care and Development*, 49–55, January. Cited on page(s) 32

Newell, A.F. and Brumfitt, P.J. (1979) Experiments concerned with the reading of Newcaster Displays. *Intl. J. of Man-Machine Studies*, **11**, 287–300. DOI: 10.1016/S0020-7373(79)80026-7 Cited on page(s) 18

Newell, A.F. and Downton, A.C. (1979) An assessment of palantype transcription as an aid for the deaf. *Intl. J. of Man-Machine Studies*, **11**, 667–680. DOI: 10.1016/S0020-7373(79)80023-1 Cited on page(s) 23, 26

Newell, A.F. and Hutt P.R. (1979) Subtitling "live" television programmes for the hearing impaired. *Intl. J. of Man-Machine Studies* **11**, 598–699. Cited on page(s) 32

Newell, A.F. (1982) Teletext for the deaf. *Electronics and Power*, **28**(23), 263–266. DOI: 10.1049/ep.1982.0112 Cited on page(s) 32

Newell, A.F. (1984) Do we know how to design communication aids? *Proc. Intl. Conf. On Rehab. Engineering,* Ottawa, Canada, 345–346, June. Cited on page(s) 39

Newell, A.F. and Brookes, C.P. (1985) Computer transcription of handwritten shorthand as an aid for the deaf–a feasibility study. *Intl. J. Man-Machine Studies,* **23**, 45–60. DOI: 10.1016/S0020-7373(85)80023-7 Cited on page(s)

Newell, A.F. (1986) Communicating via speech, the able-bodied and the disabled, (Plenary paper). *Proc. IEE Intl. Conf. On Speech Input/Output: Techniques and Applications,* London, IEE Conf. Pub. 258, 1–8, March. Cited on page(s) 108

Newell, A.F. (1987a) Computer based communication systems–the future. *Proc. 10th Annual RESNA Conference,* San Jose, California, June. Cited on page(s) 40, 75

Newell, A.F. (1987b) Technical considerations when developing a communication aid. In *Assistive communication Aids for the Speech Impaired,* P. Enderby, Ed., Pub. Churchill-Livingstone, 56–66. Cited on page(s) 77

Newell, A.F. (1988a) An idea to the market place. *Industry and Higher Education,* **2**(1), 25–28, March. Cited on page(s) 28

Newell, A.F. (1988b) A strategy for coordinated research into ordinary and extra-ordinary human computer interaction. Department of Trade and Industry, UK. Cited on page(s) 108

Newell, A.F. (1991) Assisting interaction with technology–research and practice. Invited paper at "Gladys A. Jansson Memorial Lecture to College of Speech Therapists," *British Journal of Disorders of Communication,* **26**(1), 1–10. DOI: 10.3109/13682829109011989 Cited on page(s) 39, 42

Newell, A.F., Arnott, J.L., Dye, R., and Cairns, A.Y. (1991) A full-speed listening typewriter simulation. *Intl. J. Man-Machine Studies,* **35**, 119–131. DOI: 10.1016/S0020-7373(05)80144-0 Cited on page(s)

Newell, A.F., Booth, L., and Beattie, W. (1991) Predictive text entry with PAL and children with learning difficulties. *British Journal of Educational Technology,* **22**(1), 23–40. DOI: 10.1111/j.1467-8535.1991.tb00049.x Cited on page(s) 29, 42

Newell, A.F. (1992a) Social communication: chattering, nattering and cheek. *Communication Outlook,* **14**(1), 6–8, March. Cited on page(s) 3, 50, 107

Newell, A.F. (1992b) Today's dreams–tomorrow's reality. "Phonic Ear Distinguished Lecture, AAC," *Augmentative and Alternative Communication,* **8**, 1–8, June. Cited on page(s) 72

Newell, A.F. (1992c) Whither speech systems? Some characteristics of spoken language which may effect the commercial viability of speech technology. In *Advances in Speech, Hearing and Language Processing (Vol.II),* W.A. Ainsworth, Ed., 167–197. Cited on page(s) 22

Newell, A.F., Arnott, J.L., and Waller, A. (1992) On the validity of user-modelling in augmentative communication, *Augmentative and Alternative Communication,* **8**, 89–92, June. DOI: 10.1080/07434619212331276123 Cited on page(s) 42, 79

Newell, A.F. (1993) Interfaces for the ordinary and beyond. *IEEE Software,* **10**(5), 76–78, September. DOI: 10.1109/52.232406 Cited on page(s) 46, 109

Newell, A.F. and Cairns, A.Y. (1993) Designing for extra-ordinary users. *Ergonomics in Design,* 10–16, October. DOI: 10.1177/106480469300100405 Cited on page(s) 109

Newell, A.F. (1994) Computers as better partners for everyone. Invited Keynote Address, in *Computers as Our Better Partners, Proc. of the IISF/ACM Japan Intl. Symposium,* H. Yampa, Y. Kambayashi, and S. Ohta, Eds., Tokyo, 7–13, March. Cited on page(s) 97

Newell, A.F. (1995) Extra-ordinary human-computer operation, In Extra-ordinary Human-Computer Interaction: Interfaces for users with disabilities. A.D.N. Edwards, Ed., 3–18. Cited on page(s) 109

Newell, A.F., Arnott, J.L., Cairns, A.Y., Ricketts, I.W., and Gregor, P. (1995) Intelligent systems for speech and language impaired people: a portfolio of research. In *Extra-ordinary Human-Computer Interaction: Interfaces for users with disabilities,* A.D.N. Edwards, Ed., 83–102. Cited on page(s) 40

Newell, A.F. and Gregor, P. (1997) Human computer interfaces for people with disabilities. *Handbook of Human Computer Interaction,* 813–824. DOI: 10.1016/B978-044481862-1/50101-1 Cited on page(s) 117

Newell, A.F., Langer, S., and Hickey, M. (1998) The role of natural language processing in alternative and augmentative communication. In *Natural Language Engineering,* **4**(1), 1–16, August. DOI: 10.1017/S135132499800182X Cited on page(s) 43, 64

Newell, A.F. and Gregor, P. (1999) Extra-ordinary human-machine interaction–what can be learned from people with disabilities? *Cognition Technology and Work,* **1**(2), 78–85. DOI: 10.1007/s101110050034 Cited on page(s) 109

Newell, A.F. and Gregor, P. (2002) Design for older and disabled people–where do we go from here? *Universal Access in the Information Society,* **2**(1), ISSN: 1615 5289, 3–7. DOI: 10.1007/s10209-002-0031-9 Cited on page(s) 99, 104

Newell, A.F. (2004) "HCI and older and disabled people Applied Computing, University of Dundee, Scotland," A.F. Newell and P. Gregor. *The 18th British HCI Group Annual Conference,* **2**, A. Dearden and L. Watts, Eds., Leeds Metropolitan University, UK, 6–10 September, 205–206. Cited on page(s) 6

Newell, A.F. (2006) "Older people as a focus for inclusive design." *Gerontechnology*, J.E.M.H. van Bronswijk, H. Bouma, D.G. Bouwhuis, J.L. Fozard, F.L. van Nes, and L.R. Normie, Eds., **4**(4), ISSN: 1569 1101, 190–199. DOI: 10.4017/gt.2006.04.04.003.00 Cited on page(s) 8

Newell, A.F., Gregor, P., and Alm, N. (2006) HCI for older and disabled people in the Queen Mother Research Centre at Dundee University, Scotland. Experience Report in CHI extended abstracts on Human Factors in Computing Systems, Montreal, Quebec, Canada, 22–27 April, 299–303. Cited on page(s) 120, 122

Newell, A.F., Arnott, J., Carmichael, A., and Morgan, M. (2007) Methodologies for involving older adults in the design process. *HCII. Lecture Notes in Computer Science,* Springer, 4554, Beijing, 22–27 July, **5**, 982–989. DOI: 10.1007/978-3-540-73279-2_110 Cited on page(s) 120, 121

Newell, A.F, Dickinson, A., Smith, M., and Gregor, P. (2007) "Designing a portal for older users: a case study of an industrial/academic collaboration." *ACM Trans. on Computer-Human Interactions (TOCHI),* **13**(3), 347–375. DOI: 10.1145/1183456.1183459 Cited on page(s) 84, 85, 119

Newell, A.F. (2008) "User sensitive design for older and disabled people." In *The Engineering Handbook of Smart Technology for Aging, Disability and Independence: Computer and Engineering for Design and Applications,* Helal, S., Mokhatari, M., and Abdularazak, B., Eds., Wiley, New Jersey, 787–802. Cited on page(s) 120

Newell, A.F. (2009) "Educational Videos: examining the issues older people have in using modern technology." *Interfaces 80,* Autumn 2009, 18–19. Cited on page(s) 135

Newitt, J.W. and Odarchenko, A. (1970) A structure for real time stenotype transcription. *IBM Syst. J.,* **9**, 24–35. DOI: 10.1147/sj.91.0024 Cited on page(s) 22

Newman, H. (1982) The sounds of silence in communicative encounters. *Communications Quarterly,* **30**(2), 142–149. DOI: 10.1080/01463378209369441 Cited on page(s) 48

Nielsen, J. (1993) *Usability Engineering.* London, Academic Press. Cited on page(s)

Olphert, W., Damodaran, L., Balatrsoukas, P., and Parkinson, C. (2009) Process requirements for building sustainable digital assistive technology for older people. *J. Assistive Technology,* **3**(3), 4–13, Sept. Cited on page(s) 88

Osmond, S. (1972) The Osmond report of the Lord Chancellors working party on Court Proceedings, June. Cited on page(s) 27

Park, D.C. (1992) Applied cognitive aging research. In Crail, F.I.M. and Slathouse, T.A., Eds., *The Handbook of Aging and Cognition,* Lawrence Erlbaum Ass., New Jersey. Cited on page(s) 100

Park, D. and Schwarz, N., Eds. (2000) *Cognitive Aging: A Primer.* Psychology Press, Taylor and Francis Group, Hove, 238, April 2006, 299–303. Cited on page(s) 123

Peddie, H., Cairns, A.Y., Filz, G., and Newell, A.F. (1992) A platform for extra-ordinary computer human operation (ECHO). 5th ISAAC Conference, Philadelphia. Abstract only in *Augmentative and Alternative Communication*, **8**(2), 160, August. Cited on page(s) 74

Preece, J. (1994) *A guide to usability–human factors in computing*. Addison Wesley and Open University. Cited on page(s) 118

Petrie, H. and Hamilton, F. (2004) The disability rights commission formal investigation into web site accessibility. In Dearden, A. and Watts, L. Eds., *Proceedings HCI, Design For Life*, Leeds, UK. Cited on page(s) 118, 119

Prior, S. (2010) HCI methods for including adults with disabilities in the design of CHAMPION. *Proc. CHI*, Atlanta, Georgia, 10–15 April, 2891–2894. DOI: 10.1145/1753846.1753878 Cited on page(s)

Prior, S., Waller, A., and Kroll, K. (2011) Focus groups as a requirements gathering method with adults with severe speech and physical impairments. *Behaviour and Information Technology*, (in press). DOI: 10.1080/0144929X.2011.566939 Cited on page(s) 79

Price, W.L. (1971) Palantype transcription by computer, a final report. *Report No. 5 NPL Division of Computer Science*, UK, Feb. Cited on page(s) 22

Pruitt, J. and Grudin, J. (2003) Personas: Practice and Theory. In *Proc. DUX, CD ROM*, 15 pages. DOI: 10.1145/997078.997089 Cited on page(s) 129

Pullin, G. (2007) When fashion meets discretion. Royal College of Art, 2nd–4th April. Cited on page(s) 125

Pullin, G. and Newell, A.F. (2007) "Focussing on Extra-ordinary users." *HCI. Lecture Notes in Computer Science*, Springer 4554, **5**, Beijing, 23–25 July, 253–262. DOI: 10.1007/978-3-540-73279-2_29 Cited on page(s) 126

Pullin, G. (2009) *Design Meets Disability*. MIT Press, USA. Cited on page(s) 51, 70, 71, 72, 126

Pullin, G. and Cook, A. (2010) "Insights from the six speaking chairs." *14th Biennial Conference of the International Society for Augmentative and Alternative Communication*, Barcelona, Spain, 24–29 July, 96. Cited on page(s) 70, 126

Quasthoff, U.M., and Nikolaus, K. (1982) What makes a good story–towards the production of conversational narratives. In Flammer, A. and Kintsch, W., Eds., *Discourse Processing*, North Holland, Oxford, UK, 16–28. Cited on page(s) 48, 59

Rama, M.D., Ridder, H.H., and Bouma, H. (2001) Technology generation and age in using layered user interfaces. *Gerontechnology*, **1**(1), 25–40. DOI: 10.4017/gt.2001.01.01.003.00 Cited on page(s) 103

Rau, M.T. (1993) *Coping with communication challenges in Alzheimer's disease.* Singular Publishing Group, California. Cited on page(s) 89

Reiter, E., Turner, R., Alm, N., Black, R., Dempster, M., Waller, A. (2009) Using NLG to help language-impaired users tell stories and participate in social dialogues. *Proceedings of the 12th European Workshop on Natural Language Generation,* Association for Computer Linguistics, Athens, Greece, 30–31 March, 1–8. Cited on page(s) 67

Rice, M., Newell, A.F., and Morgan, M. (2007) Forum Theatre as a requirement gathering methodology in the design of a home telecommunication system for older adults." *Behaviour and Information Technology,* 26(4), 232–331. DOI: 10.1080/01449290601177045 Cited on page(s) 89, 139

Ricketts, I.W., Cairns, A.Y., and Newell, A.F. (1995) ARCHIE–an essential component in mechatronic aids for the disabled. *Proc. of the IEE Colloquium on Mechatronic Aids for the Disabled,* Digest No. 1995/107, 11/1, May. Cited on page(s) 112

Ridgeway, L. and Mears, S. (1985) *Computer Help for Disabled People.* Souvenier Press. Cited on page(s) 16

Riley, P., Alm, N., and Newell, A.F. (2009) An interactive tool to promote musical creativity in people with dementia. *Computers in Human Behaviour,* 25, 599–608. DOI: 10.1016/j.chb.2008.08.014 Cited on page(s) 92

Roberts, S. (2010) *The Fictions, Facts, and Future of Older People and Technology,* International Longevity Centre, UK. Cited on page(s) 98

Salvador, T. and Howells, K. (1998) "Focus Troupe: using drama to create common context for new product concept end-user evaluations." *Proceedings of the Conference on CHI '98,* Summary, ACM Press, New York. DOI: 10.1145/286498.286734 Cited on page(s) 130

Sandhu, J.S. and Wood, T. (1990) Demography and market sector analyse of people with special needs in thirteen European Countries. *EEC Race 1088 Tudor Report,* Special Needs Research Unit, Newcastle upon Tyne Polytechnic, UK. Cited on page(s) 97

Sato, S. and Salvador, T. (1999) "Playacting and Focus Troupes: Theatre Techniques for creating quick, intensive, immersive and engaging focus group sessions," *Interactions,* Sept-Oct, 35–41. DOI: 10.1145/312683.312715 Cited on page(s) 130

Sayi, H., Arnott, J.L., and Newell, A.F. (1981) Stenotype-Grandjean as a speech transcription aid for the French speaking deaf. *Revue European de Diotechnologie Medicale,* 2, 465–469. Cited on page(s) 27

Schegloff, E. and Sacks, H. (1973) Opening up closings. *Semiotica,* 8, 289–327. DOI: 10.1515/semi.1973.8.4.289 Cited on page(s) 53

Scholsberg, H. (1954) Three dimensions of emotion. *Psychological Review*, **61**(2), 81–88. DOI: 10.1037/h0054570 Cited on page(s) 70

Shneiderman, B. (1983) Direct manipulation a step beyond programming languages. *IEEE Computer*, **18**(8), 57–69. DOI: 10.1109/MC.1983.1654471 Cited on page(s)

Shneiderman, B. (1986) "7 +/- 2 central issues in HCI." *Proc CHI 86, Human Factors in Computer Systems Conf.*, New York, ACM. Cited on page(s) 108

Shneiderman, B. (1992) *Designing the User Interface: Strategies for Effective Human-Computer Interaction.* Addison-Wesley Reading, Massachusetts. Cited on page(s) 118

Shneiderman, B. (1998) *Designing the User Interface. Strategies for Effective Human Computer Interaction.* 3rd ed., Addison-Wesley, 205. Cited on page(s) 44

Shneiderman, B. (2000) "Universal usability." *Communications of the ACM*, **43**(5), 7. DOI: 10.1145/332833.332843 Cited on page(s) 115

Shulman, J.D. (1979) *Captioning Reference Manual WGBH-TV,* Boston, Mass., USA. Cited on page(s) 32

Sloan, D., Ed. (2009) Special Issue on Web Accessibility Research. *Disability and Rehabilitation: Assistive Technology,* **4**(4), Informa Healthcare, 209–211, July. DOI: 10.1080/17483100902903275 Cited on page(s) 117

Sloan, D., Atkinson, M., Machin, C., and Li, K.Y. (2010) The potential of adaptive interfaces as an accessibility aid for older web users. *Proceedings of 2010 International Cross-Disciplinary Conference on Web Accessibility (W4A),* New York, ACM Press, Raleigh, US, 26–27, April. DOI: 10.1145/1805986.1806033 Cited on page(s) 105, 117, 142

Staugh, B. (2010) *The Secret Life of the Grown up Brain.* Viking Press, New York. Cited on page(s) 102

Stubbs, M. (1993) *Discourse Analysis–The Sociolinguistics of Natural Language.* Basil Blackwell, Oxford, UK. Cited on page(s) 48, 53

Sutcliffe, A. (2002) Multimedia user interface design. In *The Human Computer Interaction Handbook.* J.A. Jacko, Ed., Lawrence Erlbaum Ass., New Jersey, 245–262. Cited on page(s) 100

Swiffin, A.L., Pickering, J.P., Arnott, J.L., and Newell, A.F. (1985) PAL: An effort-efficient portable communication aid and keyboard emulator. *Proc. 8th RESNA Conf.,* Memphis, Tennessee, 197–199, June. Cited on page(s) 40

Swiffin, A.L., Arnott, J.L., and Newell, A.F. (1987) The use of syntax in a predictive communication aid for the physically handicapped. *Proc. 10th Annual RESNA Conference,* San Jose, California, 124–126, June. Cited on page(s) 43

Syme, A., Dickinson, A., Eisma, R., and Gregor, P. (2003) Looking for help? Supporting older adults' use of computer systems. In Rauterberg, M., Menozzi, M., and Wesson, J., Eds., *Human-Computer Interaction, INTERACT,* Zurich, Switzerland, 1–5 September, 924–931. Cited on page(s) 98

Taenzer, J.C. (1970) Visual word reading. *IEEE Tran. Man-Machine Systems (MMS),* **11**, 44–53. DOI: 10.1109/TMMS.1970.299961 Cited on page(s) 17

Tannen, D. (1991) *You Just Don't Understand.* Virago Press, London, UK. Cited on page(s) 52

Thompson, H. (2005) Listen-Up: Hear Wear is here. *Blueprint Magazine,* No. 232, July. Cited on page(s) 126

Todman, J. and Alm, N. (1994) Computer aided conversation: a prototype system for non-speaking people with physical disabilities. *Applied Psycholinguistics,* **115**, 45–73. DOI: 10.1017/S0142716400006974 Cited on page(s) 61

Todman, J., Elder, E., and Alm, N. (1995) Evaluations of the content of computer aided conversations. *Augmentative and Alternative Communication,* **11**, 229–234. DOI: 10.1080/07434619512331277359 Cited on page(s) 47, 61, 62, 78

Vanderheiden, G.A. (1976) Providing the child with a means to indicate. In Vanderheiden, G. and Grilley, K., Eds., *Non-vocal Communication Techniques and Aids for the Severely Physically Handicapped.* Baltimore, MD, University Park Press. Cited on page(s) 20

Vanderheiden, G.A. (1984) "High Efficiency flexible keyboard input acceleration techniques." *Proc. 2nd Ann. Conf. on Rehab Eng.,* Ottawa, June, 353–355. Cited on page(s) 39

Vanderheiden, G. and Kelso, D. (1987) Comparative analysis of fixed-vocabulary communication acceleration techniques. *Augmentative and Alternative Communication,* **3**, 181–191. DOI: 10.1080/07434618712331274519 Cited on page(s) 40

Vanderheiden, P.B. (1995) An augmentative interface based on conversational schemata. *Proc. IJCAI '95 workshop on Developing AI applications for disabled people,* Montreal, Canada, 203–212. Cited on page(s) 64

Vanderheiden, G.A. (2000) Fundamental principles and priority setting for universal usability. In Scholtz, J. and Thomas J., Eds., *Proceedings of the ACM Universal Usability Conference,* Washington, DC, 32–38. DOI: 10.1145/355460.355469 Cited on page(s) 115

Vanderheiden, G.C. (2002) A journey through early augmentative communication and computer access. *J. Rehabilitation Res. Dev.,* Nov–Dec, (6 Suppl.), 39–54. Cited on page(s) 16

VanDyke, J., McCoy, K., and Demasco, P. (1992) Using syntactic knowledge for word prediction. *ISAAC-92,* abstract in *Augmentative and Alternative Communication,* **8**. Cited on page(s) 43

Velotype (1983) *New Scientist,* 3rd., Nov. Cited on page(s)

Von Tetzchner, S. and Grove, N. (2005) *Augmentative and Alternative Communication: Developmental Issues.* John Wiley and Sons. Cited on page(s) 50

Vygotsky, L. (1962) *Thought and Language.* MIT Press, Cambridge, Mass., USA. Cited on page(s) 48

Waller, A., Beattie, W., and Newell, A.F. (1991) The computer is mightier than the sword. Newell Speech Therapy, *in Practice,* 18–20, January. Cited on page(s) 43

Waller, A., Broumley, L., Newell, A.F., and Alm, N. (1991) Predictive retrieval of conversational narratives in an augmentative communication system. *In Proc. of 14th Annual RESNA Conference "Technology for the Nineties,"* Kansas City, USA, 107–108, June. Cited on page(s) 60

Waller, A., Denis, F., Brodie, J., and Cairns, A.Y. (1998) Evaluating the use of TalksBac. A predictive communication device for non-fluent adults with aphasia. *International Journal of Language and Communication Disorders,* **33**, 45–70. DOI: 10.1080/136828298247929 Cited on page(s) 60, 65

Waller, A., Ballandin, S., O'Mara, D., and Judson, A. (2005) "Training AAC users in user-centred design." Electronic Workshops in Computing Series, *Proceedings of Accessible Design in the Digital World Conference,* 1–7, Dundee, Scotland, 23–25, August. Cited on page(s) 125

Waller, A. (2006) "Communication access to conversational narrative." *Topics in Language Disorders,* **26**(3), 221–239. DOI: 10.1097/00011363-200607000-00006 Cited on page(s) 67

Waller, A., Black, R., O'Mara, D., Pain, H., Ritchie, G., and Manurung, R. (2009) "Evaluating the STANDUP pun generating software with children with cerebral palsy." *ACM Trans. Access. Comput.,* **1**(3), 27. DOI: 10.1145/1497302.1497306 Cited on page(s) 67

Whitten, I.H., Cleary, J.G., Darragh, J.J., and Hill, D.R. (1982) Reducing keystroke counts with a predictive computer interfaces. *Proc. IEEE Computing to Aid the Handicapped,* 3–10. Cited on page(s) 40

Williamson, K., Bow, A., and Wale, K. (1997) Breaking down the barriers to public internet access. *International Conference of Computer Communication and International Telecommunications Society–joint conference,* Calgary, Alberta, Canada. Cited on page(s) 97

Wixon, D. (2003) Evaluating usability methods, interactions. *The Digital Muse,* 29–34, July/August. DOI: 10.1145/838830.838870 Cited on page(s) 120

Wright, A.G. and Newell, A.F. (1991) Computer help for poor spellers. *British Journal of Educational Technology,* **22**(2), 146–148, May. DOI: 10.1111/j.1467-8535.1991.tb00301.x Cited on page(s) 43

Yngve, V. (1970) On getting a word in edgeways. *Papers from the 6th Regional meeting of the Chicago Linguistics Society,* Chicago, USA. Cited on page(s) 53

Zebrowitz, L.A. (1990) *Social Perceptions.* OU Press, Milton Keynes, UK. Cited on page(s) 49

Author's Biography

ALAN NEWELL

Alan Newell is an Emeritus Professor at the School of Computing at Dundee University. This group was founded by Alan in 1980 and now contains one of the largest academic groups in the world whose research concentrates on computer and communication systems for older and disabled people. Alan has been researching into Human Computer Interaction, particularly by older people and people with disabilities, for over 40 years. He is a member of the Order of the British Empire and a Fellow of the British Computer Society, the Royal Society of Edinburgh and the (US) Association for Computing Machinery, and an Honorary Fellow Royal College of Speech and Language Therapy. He was a Winston Churchill Travel Fellow studying Communication Aids for the disabled in 1976 and was awarded the Lloyd of Kilgerran Prize from the UK Foundation for Science and Technology for Research into Technology to Assist People with Disabilities in 1995, the Guardian Innovation Challenge Trophy for Social Welfare 1987 with John Arnott and Norman Alm, the British Computer Society Award for Social Benefit in 1988 with Norman Alm, John Arnott and Andrew Swiffin, and the 1991 UK National Training Award for Technological Assistance for Children with Dyslexia and other Spelling and Writing Dysfunction. He was appointed a "Universal Usability Fellow" at ACM SIGCHI Conference on Universal Usability, Washington DC November 2000, and was presented with the ACM SIGCHI award for Social Impact, at CHI 2011.

Printed in the United States
by Baker & Taylor Publisher Services